Hands-On Science, Math, and Technology

Integrated Activities that Promote Critical Thinking

by David Thurlo

Good Apple
A Division of Frank Schaffer Publications, Inc.

Editorial Director: Kristin Eclov

Editor: Karen P. Hall

Illustration: Len Shelansky

Production and Design: Terry McGrath

Cover Design: Terry McGrath

Good Apple
A Division of Frank Schaffer Publications, Inc.
23740 Hawthorne Boulevard
Torrance, CA 90505

GA13018

ISBN 0-7682-0218-3

Table of Contents

3 Alternative Techniques for Measuring Distance

4 Energy and Motion

5 Creative Constructions and Contests

Introduction

Today's world is often described as a "technological society," where the skillful use of modern tools is an educational necessity for all students if they are to successfully compete in a demanding workforce as adults. To best meet the technological challenges of our society, mathematics and science should be taught as an integrated unit, rather than as two separate disciplines. Integrated lessons that combine science, mathematics, and technology not only help students understand how their skills interrelate and apply to real-world events, they also increase students' problem-solving skills and promote an active curiosity and interest in both science and mathematics.

Technology includes not just sophisticated electronic devices, but any tool that is used to measure and calculate distance, angles, and other relationships between objects. There is no doubt that computers are valuable resources when used to conduct research, write reports, and view simulations, but students are more likely to learn and remember important concepts when they experience them first-hand in a laboratory setting or out in the field.

This book offers practical, hands-on investigations that integrate concepts of earth, life, and physical science with mathematics and technology. The activities are simple, yet creative, and can be used with materials and supplies that are readily available at most schools. The following are just a few of the dynamic investigations that students plan, organize, and perform:

- Construct and use a device to measure the moon's diameter.
- Simulate the process of natural selection.
- Design and build models of towers and bridges.
- Predict and measure weather conditions.
- Use models to investigate and demonstrate radioactive decay.
- Graph and compare sunspot activity over time.
- Improve the design and performance of a paper airplane.
- Use parallax and triangulation to determine distance.

Whether you're a teacher of mathematics looking for meaningful, problem-solving activities to supplement your curriculum, or a science teacher who is in search of multi-dimensional investigations that result in objective data, the lessons in this book will provide your students with a unique, cross-curricular look at remarkable objects and events that make up our world.

Why Teach Integrated Lessons?

In schools where the curriculum allows for team or coordinated planning, there is often a need, or even a requirement, to include the interdisciplinary connection for teachers of science and mathematics who are working with the same group of students. Regardless of whether the integration is within the school, department, learning team, or just within an individual classroom, the activities taught should prove valuable to mathematics and science teachers who want their students to be active learners. Though integrating math, science, and technology may require some additional time for planning and collaboration, the benefits of teaching integrated lessons far outweigh the disadvantages. Teaching math and science in a combined curriculum

- ✦ encourages collaboration and pooling of ideas.
- ✦ allows teachers to teach their own subject matter while showing connections to others.
- ✦ improves staff morale.
- ✦ uses science to teach math and math to teach science.
- ✦ reinforces math applications and develops knowledge of science tools.
- ✦ increases interest and enthusiasm.
- ✦ reinforces learning.
- ✦ allows for activity-based strategies.
- ✦ reinforces positive peer interactions.
- ✦ enhances creativity.
- ✦ establishes career links.

How Do I Use These Activities?

The order and manner in which you choose to present the activities depends on your objectives, the needs of your students, and the requirements of your school. Before teaching any activity, look through the Teacher's Guide (pages 8–22) for suggestions on how to plan, prepare, and present the lesson. The guidelines also include ideas for follow-up activities and alternative approaches to teaching the concepts. Some activities should be taught in a specific sequence (such as the triangulation investigations), but most can be taught in any order. The contest activities (pages 21 and 22) can be taught at any time, but are especially effective as unit openers, showing students how math and science can be combined with a little bit of imagination to create a whole lot of fun!

How Do These Integrated Activities Help Students Learn?

The activities in this book help students understand the interrelationship between mathematics and science—that mathematics is necessary for the analysis and communication of scientific concepts, and through its application, provide the evidence that supports the core of major scientific theories.

In the course of investigating and analyzing mathematical and scientific concepts as they apply to technology and real-world events, students use a variety of process skills:

- ✦ Planning and conducting experiments
- ✦ Identifying and controlling variables
- ✦ Observing, comparing, and classifying properties of objects and events
- ✦ Predicting outcomes and making hypotheses
- ✦ Describing relationships, patterns, and functions
- ✦ Measuring, gathering, organizing, and interpreting data
- ✦ Estimating and computing facts and figures
- ✦ Constructing and using models
- ✦ Communicating and recording findings
- ✦ Making inferences and drawing conclusions

Teacher's Guide

Use the following suggestions to help you plan, prepare, and present each hands-on activity.

Measurement Methods and Strategies

❍ A System of Standards

Objective: Discover that measurements are inconsistent when a system of standards is not used.

Time: 45 min.

Materials needed: paper strips, pencils, metric rulers or metersticks, variety of classroom objects (textbooks, paper clips, scissors, sheets of paper, desks), calculators

Teaching tips: Invite students to discuss the difficulties and frustrations in using a mixture of units to measure and compare objects. Extend the discussion into a comparison of the various units of measure (e.g., long and short tons, metric, and SAE tools) used by different industries in countries around the world. Point out how the activity's results also help illustrate why biological scientists have turned to the binomial system of naming living things to avoid confusion when talking about an organism that has more than one common name. In addition, help students recognize that mathematics, not just science, is a standardized system, its usage of numbers and symbols in computations and formulas is basically consistent worldwide.

❍ What Unit Should I Use?

Objective: Decide which metric units are appropriate for measuring various objects. Correct answers are used to complete a crossword.

Time: 20 min.

Materials needed: dictionaries (optional)

Teaching tips: The words listed on the activity sheet are the answers to the crossword clues. Encourage students to use a dictionary to look up any of the words that are unfamiliar.

❍ Measuring Length

Objective: Measure and compare the length of various objects.

Time: 45 min.

Materials needed: metric rulers, metersticks, writing paper, graph paper, colored pencils

Teaching tips: Guide students into recognizing that metric rulers are better for measuring small items, and metersticks are better for measuring large items. Help students collect data for their graphs by setting up a data chart on the board that includes the headers "Boys Forearm Length" and "Girls Forearm Length." Have each student write his or her measurement under the appropriate header, and invite the class to refer to the listed values when constructing their graphs. Students' graphs should show forearm lengths (from shortest to longest) on the vertical axis and students' names on the horizontal axis. Remind students to use different-colored pencils to graph each set of values. Be sure students understand that the "Boys and Girls" graph will have some of the same plot points as the "Boys" graph and the "Girls" graph.

❍ Measuring Volume and Density

Objective: Devise and implement methods for determining the volume and density of objects.

Time: 45 min.

Materials needed: a variety of regular- and irregular-shaped objects (metal, plastic, and wooden blocks; solid metal cylinders; rubber stoppers; foam pieces; metal nuts and bolts; small rocks), a variety of measuring devices (triple-beam balances, metersticks, metric rulers, string, graduated cylinders, beakers), water, calculators, graph paper, reference books (math, science)

Teaching tips: Set up a rotation schedule so that each group is measuring only one item at a time. Monitor groups to make sure that all students are taking turns completing and recording the measurements. Provide students with the correct formulas needed to determine the volume of regular-shaped objects, such as a block, cylinder, and sphere. If needed, point out that the volume of an irregular-shaped object can be determined by measuring the water it displaces in a graduated cylinder. Be sure students understand that measuring an object's weight is the same as determining its mass. (Note: At a given place, equal masses experience equal gravitational forces; therefore, the masses of objects and substances may be compared by comparing the weights at the same place.)

❍ Measuring Specific Gravity

Objective: Measure and graph the specific-gravity values of various objects.

Time: 30 min.

Materials needed: solid objects (rocks, minerals, metals), spring scales, thin string or thread, scissors, large plastic beakers or containers, water, calculators, graph paper

Teaching tips: Introduce the activity by explaining that geologists use specific-gravity measurements to determine the identity of rocks and minerals they discover. Point out that the specific gravity of an object has no unit associated with it; it is simply a multiple of the specific gravity of water, which is 1. For example, quartz has a specific gravity of 2.6–2.7. Most earth science textbooks and reference materials include a list of specific-gravity measurements for common rocks and minerals. Invite students to compare their calculated values with those listed.

❍ The Importance of Significant Figures

Objective: Estimate and measure the length, mass, and volume of various solids and liquids. Then determine the number of significant figures in each measured value.

Time: 60 min.

Materials needed: metric rulers, metersticks, spring scales (that measure in grams), triple-beam balances, electronic scales (if available), small beakers (with measurement marks, so volume can be easily determined), large beakers (without measurement marks), graduated cylinders of various sizes, colored water, masking tape, permanent marker

Teaching tips: This activity requires some prep time. Each group of students will need three large, water-filled beakers labeled *1*, *2*, and *3*; three small, water-filled beakers labeled *4*, *5*, and *6*; and three different-sized, water-filled cylinders, labeled *7*, *8*, and *9*. Vary the amount of colored water in each container, and record the amounts used (so students can use these values to check their results after the investigation). If your school does not have an electronic scale, try borrowing one from a local business. Monitor students as they complete the investigation; remind them of proper safety procedures and replenish any water that has accidentally spilled. Show students how to use string to make slings for weighing objects without breakage. Note that students will be estimating, not measuring, the volumes in beakers 1, 2, and 3 since those beakers have no measurement markings. If students are unclear how to determine the number of significant figures, show them one or two sample calculations. For example, 24.0 cm has three significant figures. Math teachers may wish to go into a more in-depth explanation of significant figures, including a guided lesson with practice problems on the board.

❍ Devising Measuring Strategies

Objective: Devise and implement indirect methods for measuring tiny, irregular-shaped objects.

Time: 2-day activity, 30 min. each day

Materials needed: various measuring tools that students request, such as medicine droppers, calculators, hand lenses, micrometers, calipers, and an electronic scale (one that accurately measures powders)

Teaching tips: Encourage students to assist you in collecting the needed items (determined during the planning phase of this activity, day 1). Guide students in carrying out their strategies. If needed, provide some possible strategies: determine the average length or mass of a small item by weighing a specific number of the items and then dividing the total weight by the number of items measured; and determine the area of a metal eraser holder by multiplying the length and width of a paper strip that was wrapped around the pencil and cut to match the dimensions of the eraser holder. Be sure to discuss the potential errors introduced with each measurement method used.

❍ Do You Overestimate or Underestimate Yourself?

Objective: Calculate the difference between estimated values and actual measurements of body parts.

Time: 30 min.

Materials needed: metersticks, string, tape measures

Teaching tips: Review how to add positive and negative numbers. Encourage students to seek help from others when measuring challenging areas, such as the circumference of their finger or head.

Exploring the Environment

❍ Determining the Moon's Diameter

Objective: Construct and use a device for measuring the diameter of a full moon.

Time: 30 min. in-class prep, 30 min. homework, 20 min. wrap-up

Materials needed: metersticks, 3" x 5" (7.5 cm x 12.5 cm) index cards (one per student), 5" x 7" (12.5 cm x 17.5 cm) index cards (one per student), X-acto® knives or single-edged blades (teacher-supervised use), pens, aluminum foil, masking tape, straight pins

Teaching tips: This activity needs to be done at night during a full moon, so plan ahead by checking in your local newspaper or wall calendar to track the lunar cycle. Help students prepare their measuring devices in class, and then invite students to use their devices at night to measure the diameter of a full moon. Encourage family involvement. The next class session, have students share their results, and tell them the actual diameter of the moon (3,476 km) so that they can complete the rest of the activity.

❍ Scale Models of Planets and Moons

Objective: Draw scale models that compare the diameters of planets and moons.

Time: 60 min.

Materials needed: 4-m strips of adding-machine tape, metersticks, pencils

Teaching tips: In advance, have student volunteers help cut adding-machine tape into 4-m strips. You may choose to post pictures of celestial bodies (e.g., planets, stars, and the moon) to motivate learning. Introduce the activity by inviting students to share their experiences constructing scale models, such as airplanes, ships, and buildings. Draw a simple sketch of the activity setup to help illustrate how the celestial bodies are being compared, and discuss why a strip of adding-machine paper will serve the purpose better than a 4-m square of paper. Provide additional 4-m strips to students who wish to complete the extension activity (making scale models to show the distance from the sun to various planets). Follow-up this investigation with other related activities: invite students to make circular scale models of the planets to post in the classroom. To further explore the concept of drawing models to scale, have students make 1:1 scale drawings of animals by tracing onto butcher paper projected images that have been adjusted to match the actual sizes of the animals.

❍ Star Light, Star Bright

Objective: Draw a graph to compare the luminosity and surface temperatures of various stars.

Time: 45 min.

Materials needed: metric rulers, colored pencils, Star Chart (page 51) photocopies, Hertzsprung-Russell Diagram (page 52) photocopies

Teaching tips: Invite students to recall their experiences star gazing. Point out that the brightness of a star viewed from Earth is a result of the star's size, distance, and temperature. Before having students complete the activity independently, demonstrate how to plot points on the Herzsprung-Russell Diagram. As students complete the chart, they will notice that most of the stars appear in a loose diagonal band known as the "main sequence." The few stars plotted in the upper right are big, cool stars called *red giants*, and the ones in the bottom left are small, hot stars called *white dwarfs*. Let students know that the ultimate fate of a star depends on its original size; the smaller stars have much longer lives. A good follow-up to this activity would be to investigate a star's life cycle, which includes black holes, neutron stars, pulsars, and black dwarfs. Enhance learning by posting star charts and showing videos about the stars.

❍ Predicting Sunspots

Objective: Graph and compare data that summarizes sunspot activity over a 100-year time period.

Time: 45 min.

Materials needed: graph paper, calculators, Sunspots Table (page 55) photocopies

Teaching tips: Sunspots have been observed for hundreds of years, and may be viewed indirectly by aiming a small telescope or high-powered binoculars at the sun and then focusing the sun's image on a piece of paper. CAUTION: Never allow students to look directly at the sun. A cardboard collar around the telescope or binoculars helps insure that only light going through the lenses is focused on the paper. The rotational period of the sun may be timed by following the motion of sunspots across the sun's surface.

❍ How Much Humidity?

Objective: Measure the amount of humidity in the air.

Time: 30 min.

Materials needed: Celsius thermometers (2 per student pair); white, wide, hollow shoestrings; white thread; scissors; small beakers of water; Relative-Humidity Chart (page 58) photocopies

Teaching tips: This activity should precede the Weather Watch activity (page 59). In advance, cut shoe strings into halves or fourths for students to use. Note that gauze may be substituted for shoe string. If supplies are limited, three or four pairs can share one beaker of water. Be certain that Celsius thermometers are the same design (one is not smaller or labeled differently) so that students can accurately compare the different temperature readings. Discuss the meaning of humidity in the amount of water vapor in the air. Invite students to explain why the rate of evaporation is affected by humidity. Explain that we feel hot and sticky on humid days because the air is too saturated to absorb the perspiration from our skin. A good follow-up to this activity is to have students research and report on how various air conditioners work.

❍ Weather Watch

Objective: Use various tools to observe, measure, and compare weather conditions.

Time: 1-month activity: 30 min. the first day, 10 min. each subsequent day, 30 min. wrap-up

Materials needed: thermometers (Fahrenheit, if possible; two per student group), standard aneroid barometers, wind meters, weather vanes, Relative-Humidity Chart (page 58) photocopies, Weather Watch Chart (page 61) photocopies, photographs of cloud types (or reference books)

Teaching tips: This activity should be done after completing How Much Humidity? (page 56). Have students make wet- and dry-bulb thermometers by following the procedure they used in that activity. Show students how to use all the weather equipment before they begin the investigation. Ideally, students should take weather readings every day for one month. If supplies are limited (for example, barometers are unavailable), have students call the weather bureau to obtain readings or have them listen to a weather radio station. Computers with modems can also be used to obtain weather readings from nearby weather stations. Check with the weather department of a local television station. Many times these stations will provide additional resources for students to use in weather-related activities.

❍ Survival of the Fittest

Objective: Calculate the survival rates of different-sized "rodents" in a natural-selection simulation.

Time: 45 min.

Materials needed: colored construction paper, metric rulers, scissors, large containers (1 per student group), calculators

Teaching tips: Topics to discuss along with this activity could include population sampling, probability studies, food webs, predator-prey relationships, extinction, and Darwin's theory of natural selection. To extend learning, have students graph their results.

❍ A Salty Solution

Objective: Compare the mass and volume of a saltwater solution to the mass and volume of its constituents.

Time: 30 min.

Materials needed (per group): 100-mL graduated cylinder, 150-mL beaker, triple-beam balance or electronic scale, cup of salt, cup of water

Teaching tips: Ask students to explain why the volume measurements don't add up and to infer where the extra volume of matter has gone. Extend the activity to include comparing the results of dissolving other materials in water, such as instant coffee and sugar, or comparing the volume and mass of a carbonated drink before and after the carbon dioxide has escaped. Introduce the terms *solute* (the substance being dissolved) and *solvent* (the liquid in which the solute is dissolved), and then invite students to think of other examples of solutes and solvents.

Alternative Techniques for Measuring Distance

❍ The Parallax Principle

Objective: Use the phenomena of parallax to help determine the distance to a drawn line.

Time: 45 min.

Materials needed: chalkboard or posted butcher paper, chalk or markers, metric rulers, metersticks, masking tape, calculators

Teaching tips: If chalkboard space is limited, post long strips of butcher paper on the classroom walls for students to draw on. Help students decide where to stand (in front of the posted paper) to avoid congestion. Students should be able to determine why they should make their initial line on the board (or butcher paper) directly in front of their sighting position (because angles create lopsided triangles). Encourage students to practice the technique of parallax a few times before performing the actual test. Extend learning by discussing how range-finders, largely replaced by the end of WWII by radar (and recently lasers), use parallax to determine the distance to targets.

❍ Angular Measurements

Objective: Measure angles by using a protractor and a magnetic compass.

Time: 30 min.

Materials needed: protractors, magnetic compasses, metric rulers

Teaching tips: Use this activity to prepare students for the triangulation activities. Be certain students know how to properly use protractors; they should place the baseline of the protractor parallel to the line of one of the angles, and position the crosshairs or reference point at the apex of the angle being measured. Simple compasses are okay to use for this activity, but check to make sure that magnets and large, metal objects do not affect them. Compasses that show degree markings as well as direction labels are much easier for students to use. Discussions that could accompany this activity include how compasses work, the difference between magnetic north and the geographic North Pole, and the sport of orienteering (in which participants use compasses and outdoor skills to follow a set course).

❍ Mapping and Triangulation

Objective: Measure the angles of triangles by using a protractor and a calculator.

Time: 30 min.

Materials needed: protractors, calculators

Teaching tips: This activity expands on the skills practiced in Angular Measurements (page 70) and prepares students for Triangulation on a Scale Map (page 74). Again, be certain that students know how to properly place and use a protractor when measuring angles.

❍ Triangulation on a Scale Map

Objective: Use the technique of triangulation to calculate distances on a scale map, and then convert those scale measurements to actual lengths.

Time: 30 min.

Materials needed: protractors, metric rulers

Teaching tips: This activity builds on the skills used in Angular Measurements (page 70) and Mapping and Triangulation (page 72). At the end of the activity, have students compare answers, and point out the importance of making careful, accurate measurements when using triangulation to draw scale maps. Note that the last question of the "Follow-Up" section, using triangulation to determine the height of a flagpole, is a tie-in to the activity Using Right Triangles to Measure Height (page 77).

❍ Using Right Triangles to Measure Height

Objective: Construct and use a triangular-shaped measuring device to determine the height of a flagpole.

Time: 45 min.

Materials needed: 5" x 7" (12.5 cm x 17.5 cm) index cards, scissors, calculators, metersticks, tape, small spirit levels

Teaching tips: This activity should be done after Triangulation on a Scale Map (page 74). If spirit levels are unavailable, a string and weight may be attached to the triangle as a plum bob, or partners may visually insure that the base of the triangle is parallel to the ground. To help students visualize why the height of the flagpole is $h + d$, draw a sketch on the board or provide an example drawn on graph paper.

❍ Triangulation in Real Life

Objective: Use the technique of triangulation to determine the distance to objects in the classroom.

Time: 45 min.

Materials needed: adding-machine tape or butcher paper, metersticks, protractors, straws, graph paper

Teaching tips: Prior to beginning the activity, tape adding-machine tape or butcher paper to students' desktops. Select stationary objects in the classroom, such as a fire extinguisher or a poster, for students to use in this activity. Objects at the same height as a desk or table are easiest for students to observe. Note that results will vary, depending on where students measure from. At the end of the activity, have students check the accuracy of their calculations by using metersticks to measure the actual distance to each object. You might also discuss the advantages and disadvantages of using very large baselines. This activity can be conducted outside if students have sighting surfaces to work from. To extend learning, invite an expert to demonstrate how survey instruments are used to measure distances and elevations.

Energy and Motion

❍ Mathematical Model of Radioactive Decay

Objective: Simulate radioactive decay, and calculate the decay rate of "atoms."

Time: 60 min.

Materials needed (per student pair): cup of uncooked rice, markers, shoe box or other four-sided box with lid, masking tape, graph paper

Teaching tips: Unpopped popcorn may be substituted for the rice, with the "pointed end" of the kernel being used in place of the colored tip. As a fun introduction to this activity, use popping corn kernels to illustrate a chain reaction. In advance, pop some popcorn in a microwave oven, and make a tape-recording of the sounds. Students can listen to the tape, note the time between each kernel pop, and make a graph of the results (Number of Pops vs. Time).

❍ Melting Ice

Objective: Measure and graph the temperature changes of melting ice.

Time: 45 min.

Materials needed: ice chest filled with ice cubes, foam cups, water at room temp., Celsius thermometers (1 per group), clock with a second hand, sponges or paper towels, graph paper

Teaching tips: As a variation of this activity, students can test salt water instead of or in addition to regular water. You may either inform students of what is being tested, or students can determine that the samples are different by comparing the results.

❍ Vapor Escaper

Objective: Determine the relationship between surface area and evaporation rate.

Time: 4-day activity: 20 min. the first day, 10 min. each subsequent day, 30 min. wrap-up

Materials needed (per group): 3 different-sized, cylindrical, empty, metal food cans; masking tape; marker; metric ruler or calipers; water; triple-beam balance or electronic scale; graph paper

Teaching tips: Use cylindrical containers that vary significantly in size, such as one-gallon (4.5-L) cans, soup or stew cans, and small tomato-paste cans. Remind students to check their setups to be sure that all three cans are exposed to the same conditions. If weather conditions are unfavorable, such as high humidity, this experiment may require extra time. As long as all three cans are exposed to the same conditions, the duration of the test shouldn't affect the results. The rate of evaporation of water from cylindrical containers should be proportional to the exposed surface of the water if no other variables are introduced (e.g., one can is placed in the shade while the other two are not).

❍ Turn Up the Heat

Objective: Measure, graph, and compare the heating and cooling rates of different-colored surfaces.

Time: 60 min.

Materials needed (per group): 2 different-colored sheets of construction paper, foam board or cardboard, 4 rubber bands, 2 Celsius thermometers, ring stand, 100-watt lightbulb in a clamp-light, scissors, metric ruler, protractor, tape, calculator, writing paper, graph paper, colored pencils, textbooks

Teaching tips: Assign different colors to each group so that a variety of colors are tested. Caution students about touching the clamp-light's hood and the bulb, and tell students to stay clear of the electrical cords. To help students collect class data, have each group write their results on the chalkboard. As an alternative approach to this activity, students can take temperature readings outside and use the sun as a heat source. They should assemble the apparatus indoors and then block out the sun when taking the initial reading outdoors. Have them use wristwatches or stopwatches to keep track of the time. Tell students to be sure not to cast shadows on the thermometers when taking the temperature readings.

❍ Stretching the Truth

Objective: Measure and compare how much a rubber band stretches when increasing amounts of weight are applied.

Time: 30 min.

Materials needed: metersticks, large paper clips, string, washers (12 per student), graph paper, scissors, rubber bands

Teaching tips: Make sure the washers used for this activity are all identical. Check students' setups and adjust them, if necessary, so that they work properly. Make sure the metersticks are secure, not on the edges of the desks, so that the weight of the washers doesn't bend the wood and affect the results. Note that small springs may be used instead of rubber bands. This activity can be used to reinforce lessons about linear functions. You may also choose to tie this activity into lessons about other related math concepts, such as direct proportions, inverse relationships, and quadratics.

❍ Follow the Bouncing Ball

Objective: Construct and use a "bounce chamber" to measure the height and frequency of a bouncing ball.

Time: 2-day activity: 60 min. for the first day (construction and testing) and 45 min. for the second

Materials needed (per group): 3 different types of balls, string, meterstick, stiff poster board, pencil, scissors, markers, clear plastic wrap, clear tape, video equipment (video camera on a tripod, videocassette, VCR connected to a TV), graph paper

Teaching tips: This activity can also be done as a 3-day activity: day 1, construction of the bounce chamber; day 2, testing; and day 3, analyzing and summarizing data. Monitor students' progress as they build the bounce chambers, and assist them if needed. Note that the bounce chamber keeps the bouncing ball from taking on a lateral movement, thus allowing for more accurate measuring of height. Provide balls that are easily tracked against the poster board backing, such as tennis balls, Ping-Pong™ balls, handballs, and other high-bouncing balls. The chamber should be placed on uncarpeted floor, and should be held in place to insure that friction from the sides of the chamber do not alter the results. Since each group will need to use a video camera, set up a rotation schedule and ask students waiting for a turn to complete another assignment. You may choose to have each group bring in their own videotape to avoid possible mix-ups and accidental erasing. If a video camera and VCR are not available, students will need to keep a sharp eye on the bouncing ball and most likely will require multiple trials in order to gather the necessary data.

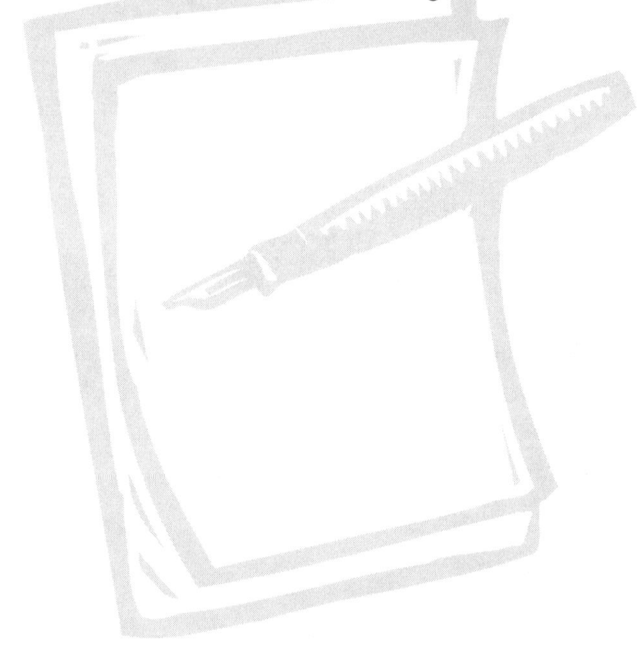

Creative Constructions and Contests

❍ The Tallest Tower

Objective: Design and construct a tall tower that can withstand strong winds.

Time: 45 min.

Materials needed (per student pair): 25 straws, 100 straight pins, electric fan (1 per class)

Teaching tips: Prior to beginning this activity, encourage students to research and report on various building designs, and invite a civil engineer to speak to the class about his or her experiences. Monitor students as they construct their towers, cautioning them about the proper handling of pins. To ensure that the competition is fair, quickly inspect each tower before placing it in front of the fan. Also, use a triple-beam balance to quickly check whether rules concerning the number of straws and pins were followed: record the weight of 25 straws and 100 pins, and then weigh each tower; any structure exceeding the weight of the materials is not allowed in the contest. Test the towers on carpeted floor (make sure that no pins are anchoring the towers to the carpet). Test one tower at a time, placing it in front of the fan from a set distance. The towers that remain intact and standing are put through subsequent rounds of testing in which they are closer to the fan or exposed to higher speeds.

❍ The Strongest Bridge

Objective: Design and construct a durable bridge that can withstand added weight.

Time: 45 min.

Materials needed (per student pair): 25 straws, 100 straight pins, metric ruler, string, standard weights (1 set per class), stopwatch (1 per class), pillow (1 per class)

Teaching tips: Use any standard set of weights, such as those used in balances. Attach paper-clip hooks to the weights so they can be hung easily from each bridge. If standard weights are not available, use metal buckets made from various-sized cans (with wire bails) and add measured amounts of sand to the buckets to increase the weight. As students build their bridges, move two tables to within 20 cm of each other. When testing the bridges, let students select how much weight to add after each round of testing. Be sure to place a pillow under the tested bridge to catch the falling weights and to avoid possible damage to the floor.

❍ Oops! The Egg Dropped

Objective: Design and construct protective packaging for an egg to keep it from cracking when dropped.

Time: 45 min.

Materials needed: 1 raw chicken egg (per student), 25 drinking straws (per student), 100 straight pins (per student), tall extension ladder, plastic drop cloth, lined trash can, spatula, dustpan

Teaching tips: Before beginning the testing, cover the ground with a plastic drop cloth, and place clean-up supplies (lined trash can, spatula, dustpan) nearby. For safety reasons, an adult should be responsible for climbing the ladder and dropping the containers. You may choose to drop the containers from an elevated area of the school, such as the gymnasium roof. While testing, have students stand far away from the drop zone to avoid being stuck by pin-studded containers or slipping on broken eggs.

❍ First-Rate Flying

Objective: Design and construct a paper airplane that is capable of flying far distances for long periods of time.

Time: 45 min.

Materials needed: sheets of white copy paper, scissors, tape measure or meterstick, stopwatch

Teaching tips: If time permits, students should test their airplanes two or three times and then take the average of the measurements. If time is limited, have three or four students line up at the launch site (with plenty of space in between them) and propel their airplanes simultaneously. Recruit the help of volunteers, and provide extra stopwatches and measuring tapes or metersticks for students to use to measure the duration and distance of each flight. You may choose to vary this activity by permitting students to select additional building materials (such as glue, tape, paper clips, and craft sticks), and have a "free form" flying contest.

Name ______________________ Date ____________

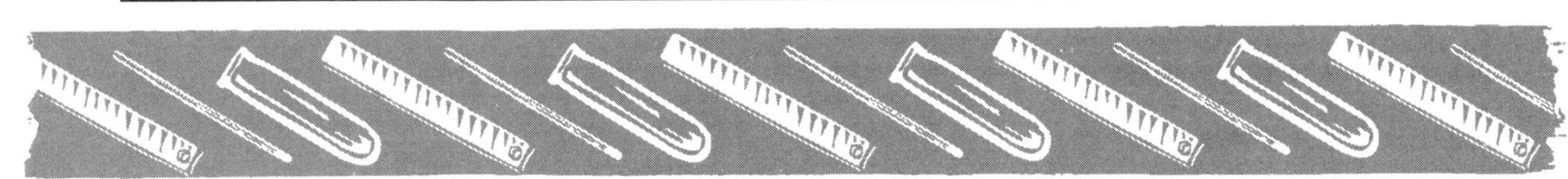

A System of Standards

Purpose: Identify the difficulties in comparing measurements when standard units are not used.

Background: Throughout history, a variety of units have been developed and used for measuring. For example, feet, inches, yards, meters, and leagues are all units of length. The International System of Units (SI) is the set of standard measurements used by mathematicians and scientists worldwide. By using these standard measurements, mathematicians and scientists are able to communicate and compare their results much more easily and accurately.

Materials:

- paper strips
- pencil
- items to measure (textbook, paper clip, sheet of writing paper, scissors, a desk)
- metric ruler or meterstick
- calculator (optional)

Procedure: Read the entire procedure before you begin the activity.

1. On a strip of paper, mark with a pencil the maximum width of your index finger. This width will be your "small personal unit of length," which will be called a *nub*. Make a paper ruler that is ten nubs long.
2. On another strip of paper, mark the length from your elbow to your wrist. This distance will be your "large personal unit of length," which will be called a *stem*. Fold the paper strip in half lengthwise, and then fold it in half again to make four sections of equal length. With a pencil, mark off ¼, ½, and ¾ of a stem to complete this second paper ruler.
3. Complete Part I of the "Data and Results" section, using your paper rulers and a metric ruler or meterstick to measure the listed items.
4. Use your results from Part I to complete Part II of the "Data and Results" section.
5. Compare your completed charts to those of your classmates. Then answer the follow-up questions.

Name ______________________________ Date ____________

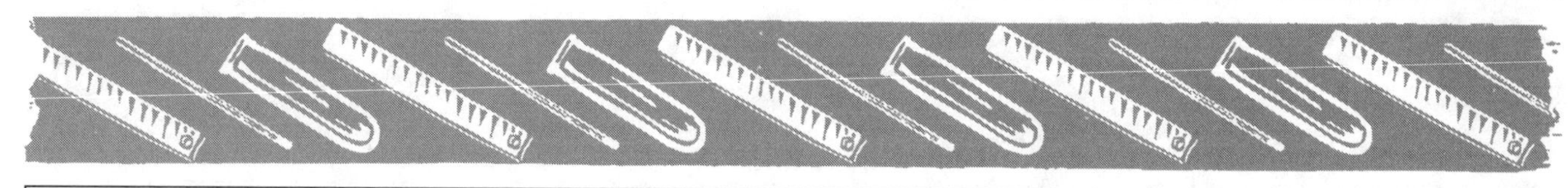

A System of Standards (continued)

Data and Results

Part I. Use your paper rulers and a meterstick or metric ruler to complete the following measurements.

TABLE A

Item	Value in Nubs	Value in Stems	Value in Centimeters
textbook length			
textbook width			
textbook thickness			
paper-clip length			
scissors length			
paper length			
paper width			
desktop length			
desktop width			
desktop thickness			

Part II. Convert your nub and stem measurements into centimeter equivalents, and list the values in Table B. To determine the conversion ratios, align a metric ruler next to each paper ruler to see how many centimeters equal one nub or one stem.

Name ______________________________ Date ____________

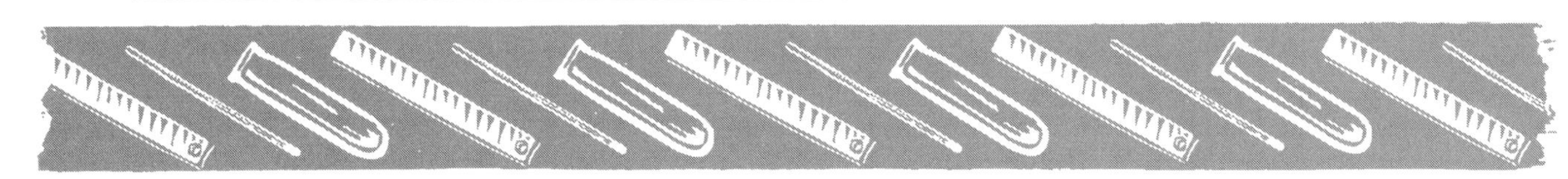

A System of Standards (continued)

TABLE B

Items	Nubs → Centimeters	Stems → Centimeters
textbook length		
textbook width		
textbook thickness		
paper-clip length		
scissors length		
paper length		
paper width		
desktop length		
desktop width		
desktop thickness		

Follow-Up

1. Compare the converted values to the calculated centimeter values in Table A. Which was the more accurate unit of measurement, the nub or the stem? Explain your answer.

____

____

____

2. Are your nub and stem measurements identical to those of your classmates? Why?

____

____

3. Why is it important for scientists and mathematicians to use standard units rather than a personal system to measure items?

____

____

Name ______________________________ Date ______________

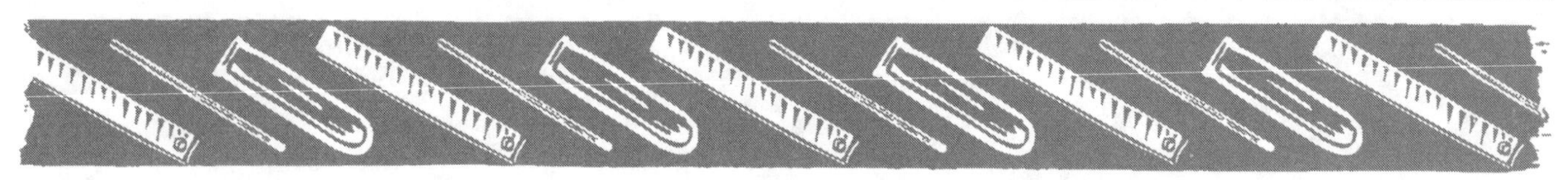

What Unit Should I Use?

Purpose: Determine the appropriate unit of measure for selected objects.

Background: The choice of which unit of measure to use depends on the size of what you're measuring and whether you're measuring length, area, volume, or weight. A smaller unit will give you a more exact measurement, but its use may not always be practical.

Materials:

dictionary (optional)

Procedure: Read the clues and use the words from the word list to complete the following crossword.

Word List

angstrom
centimeter
gram
kilogram
kilometer
liter
micrometer
milligram
milliliter
millimeter

ACROSS

2. the amount of potato chips in a bag
4. the length of a blood cell
5. the volume of water in a test tube
6. the length of the Rio Grande
7. the amount of cola in a large plastic container
8. the amount of sugar in a piece of gum

DOWN

1. the diagonal of a computer monitor screen
3. the diameter of an oxygen atom
5. the width of your thumbnail
6. the weight of a dog

Name ______________________________ Date ______________

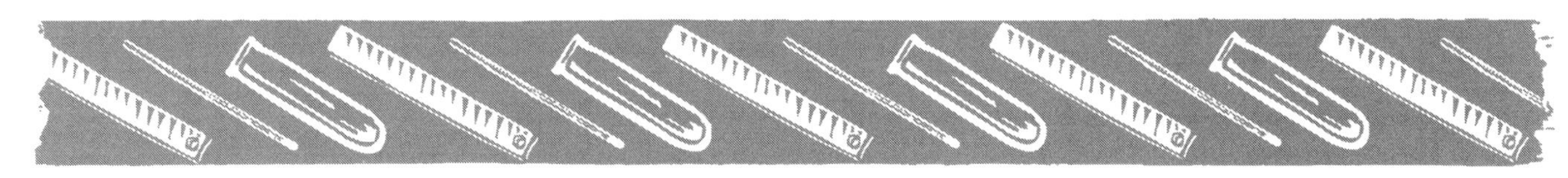

Measuring Length

Purpose: Measure length in millimeters, centimeters, and decimeters.

Background: The metric system is an international system of weights and measures based on the decimal system. It was devised in 1791 by the French Academy of Sciences, and was standardized in 1875 by the International Bureau of Weights and Measures.

Materials:

- metric ruler
- meterstick
- writing paper and graph paper
- colored pencils

Procedure: Measure the length of each line in decimeters, centimeters, and millimeters. Record your measurements in the Answer Key of the "Data and Results" section. Then complete the "More Measurements" section.

1. ________________
2. ______________
3. ___________________________
4. ____
5. _____________________
6. ______________________________________
7. ____
8. ____
9. _______
10. ________________
11. __________________________________
12. ____
13. ____
14. ____
15. ____

Name ______________________________ Date __________

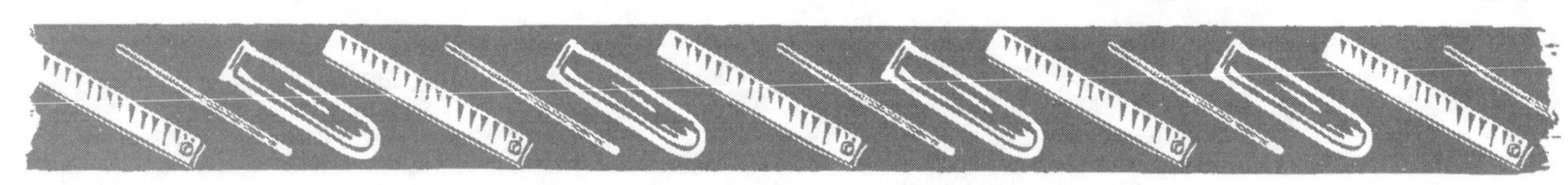

Measuring Length (continued)

Data and Results

ANSWER KEY

1. ____________	6. ____________	11. ____________
2. ____________	7. ____________	12. ____________
3. ____________	8. ____________	13. ____________
4. ____________	9. ____________	14. ____________
5. ____________	10. ____________	15. ____________

More Measurements

1. The length of this paper is __________dm.
2. The thickness of your math textbook is __________mm.
3. The length of your desk or tabletop is __________cm.
4. The diagonal distance across your desk or tabletop is __________cm.
5. The volume of your textbook is __________cm^3.
6. The classroom door is __________cm high and __________cm wide.
7. The chalkboard is __________m long and __________m from the floor.
8. The width of the trim around the chalkboard is __________mm.
9. The length of your forearm, from your elbow to the tip of your middle finger, is __________cm.
10. On a separate sheet of paper, record your classmates' answers for #9. Then graph the following on one sheet of graph paper. Use different-colored pencils to draw the three line graphs, and include a color key to identify each one.
    - Boys' forearm length from shortest to longest
    - Girls' forearm length from shortest to longest
    - Combined boys' and girls' forearm lengths from shortest to longest

Name ______________________________ Date ______________

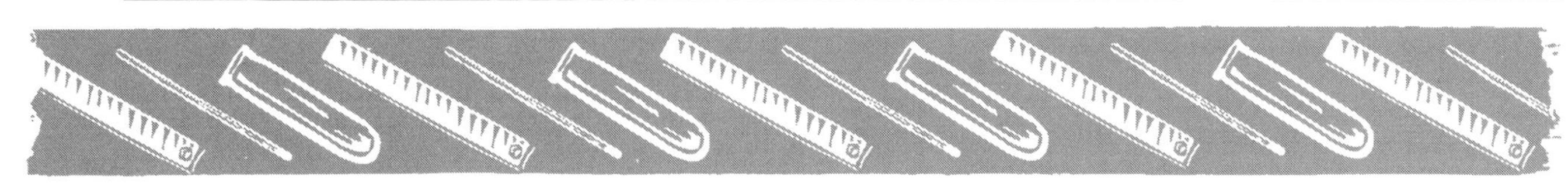

Measuring Volume and Density

Purpose: Plan and implement ways to measure the volume and density of various objects.

Background: The volume of an object is the amount of space it occupies. The density is the mass per unit volume (d = m/v).

Materials:

- various substances (metal, plastic, and wooden blocks; solid metal cylinders; rubber stoppers; foam pieces; metal nuts and bolts; small rocks)
- measuring devices (triple-beam balance, meterstick, metric ruler, string, graduated cylinders, beakers)
- water
- calculators
- reference books (math, science)
- graph paper

Procedure: Work with two classmates. Read the entire procedure before beginning the investigation. Be sure to share supplies with other groups. Use only one item at a time, and return the item to the supply area when you are finished.

1. Choose six items from the supply your teacher provides. (The objects you choose should vary in size and shape.) List these items in the Measurements Table of the "Data and Results" section.
2. As a group, plan how you will measure the volume and density of each object. Brainstorm the equipment you will use, the steps you will take, and the formulas you will need to complete each measurement. Write your plan in the "Data and Results" section.
3. Gather the measurement devices and reference books you need to complete the investigation. Then carry out your plan and complete the Measurements Table.
4. Draw a bar graph of the densities you calculated. Graph the values from the most to the least dense. List on the x-axis (horizontal axis) the names of the items tested, and list on the y-axis (vertical axis) a density scale (from 0 to the highest density you measured).

Name ______________________________ Date ______________

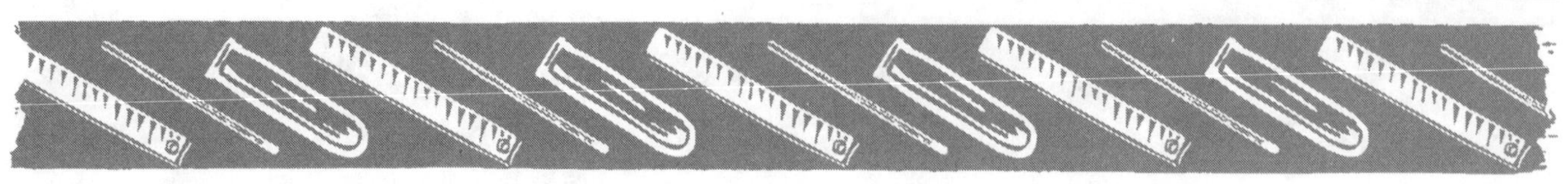

Measuring Volume and Density (continued)

Data and Results

Our Plan:

____
____
____
____
____
____
____
____
____

MEASUREMENTS TABLE

	Item	Mass	Volume	Density
1.				
2.				
3.				
4.				
5.				
6.				

Name ______________________________ Date ______________

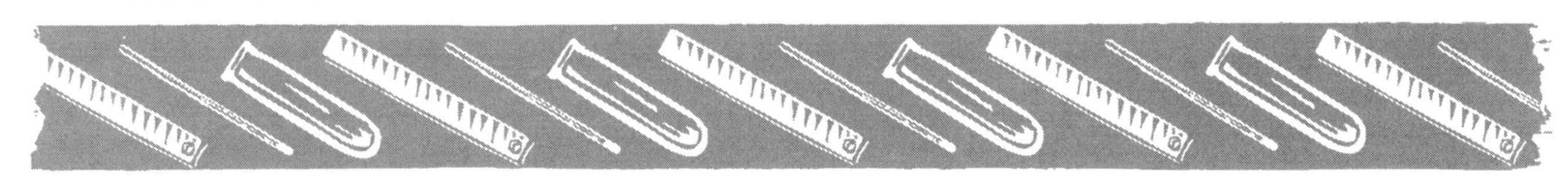

Measuring Volume and Density (continued)

Follow-Up

Discuss your results with classmates. Then use the information you learned from the investigation, as well as your prior knowledge of volume and density, to answer the following questions. You may work independently or with your group.

1. What is density?

 ____

 ____

2. What two measurements do you need in order to calculate the density of an object?

 ____

 ____

3. How can you determine the mass of a solid object? How can you determine the mass of a liquid?

 ____

 ____

 ____

 ____

4. How can you determine the volume of a solid with regular dimensions, such as a cube or a cylinder? How can you determine the volume of a liquid?

 ____

 ____

 ____

 ____

5. How can you use water to determine the volume of an irregular-shaped object?

 ____

 ____

 ____

Name ______________________________ Date ______________

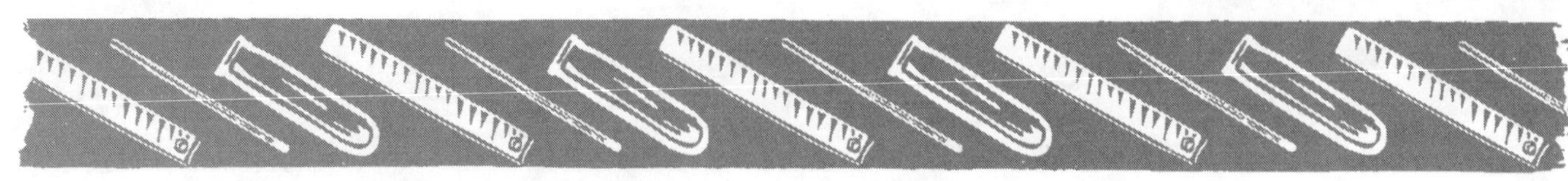

Measuring Specific Gravity

Purpose: Determine the specific gravity of selected objects.

Background: *Hefting* is the process of lifting and handling two objects to compare their weights. A more accurate way to compare solids, such as rocks, minerals, and metals, is by their relative densities or specific gravity. The specific gravity of a substance is its density relative to (divided by) the density of an equal volume of water.

Materials

- 6 solid samples (rocks, minerals, metals)
- spring scale
- thin string or thread
- scissors
- large plastic beaker or container
- water
- calculator
- graph paper

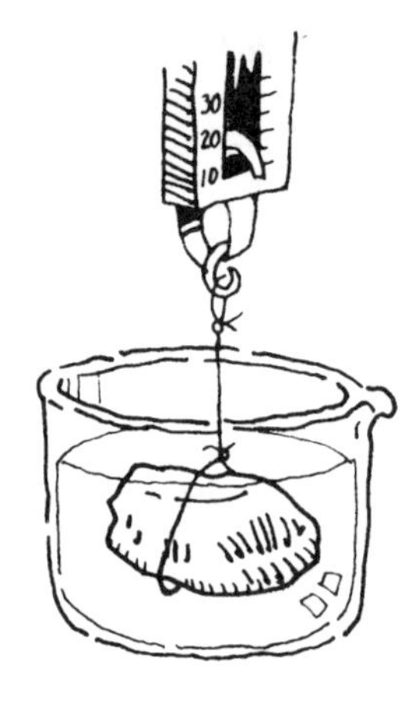

Procedure: Choose one or two classmates to work with. Read the entire procedure before you begin the activity.

1. List in the Measurements Table of the "Data and Results" section the solid samples you will be measuring.
2. Tie a string around one of the solids, and hang it on the spring scale. Read the solid's mass, and then record the measurement (in grams) in the data table under the header *Mass in Air.*
3. With it still hanging from the string, place the solid into a beaker ¾ full of water. The solid should be completely submerged, but not touching the sides or the bottom of the beaker. Measure the solid's mass again, and record the value on Table A under the header *Mass in Water.*
4. Repeat steps 2 and 3 for each of the remaining solids. Then calculate the specific gravity of each sample by using the following formula. To determine the mass of an equal volume of water, subtract the sample's weight in water from the sample's weight in air. Record specific-gravity values in the table.

$$\text{Specific gravity} = \frac{\text{Mass of the sample in air}}{\text{Mass of an equal volume of water}}$$

5. Draw a bar graph that compares the specific-gravity values of the samples. List on the x-axis the names of the samples, and list on the y-axis the specific-gravity values.

Name ____________________ Date __________

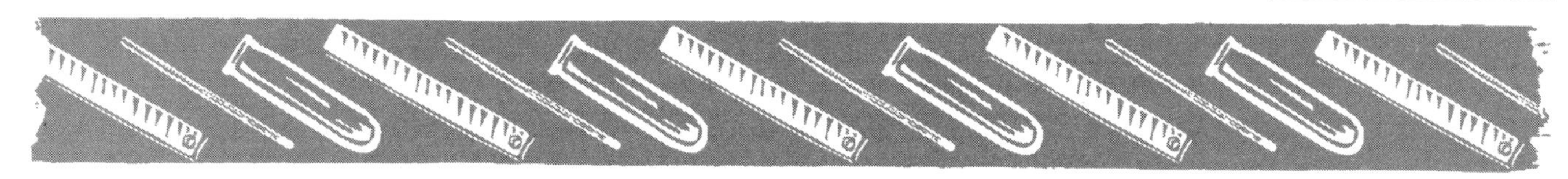

Measuring Specific Gravity (continued)

Data and Results

MEASUREMENTS TABLE

	Sample	Mass in Air	Mass in Water	Specific Gravity
1.				
2.				
3.				
4.				
5.				
6.				

1. Based on your results, rank the samples from lowest to highest specific gravity.

   ____

   ____

2. Compare your results with those of your classmates. Which sample was most often listed as the highest specific gravity? the lowest specific gravity?

   ____

   ____

3. Using the class data, calculate the average specific gravity of each sample. Then rank the average values from lowest to highest. How does this ranking compare to the one done with your results?

   ____

   ____

   ____

4. What possible errors could have affected the results of this activity?

   ____

   ____

Name ______________________________ Date ______________

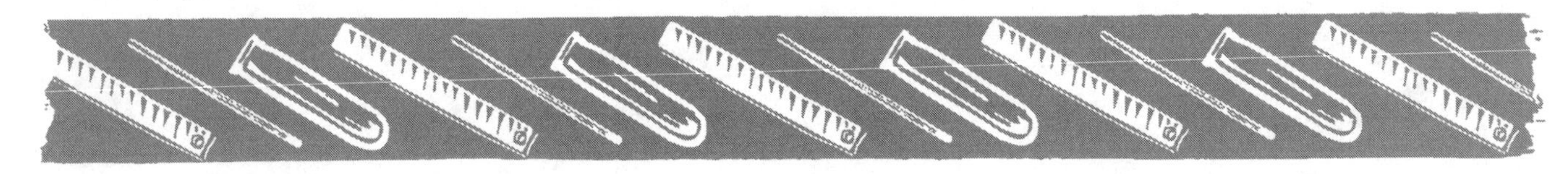

The Importance of Significant Figures

Purpose: Use estimation, approximation, and precision to enhance measuring skills and to understand the importance of significant figures.

Background: Every measurement is an approximation of the best value possible with the equipment being used. A person standing on a truck scale may note an estimation of their weight as 100 lbs., but the same person on a more precise doctor's scale may find their actual weight to be 108.5 lbs. The precision of a measurement is determined by how many digits are obtained by actual measurement. Each of these digits is called a *significant figure*. The number of significant figures in a measurement indicates how precise the measurement is.

Materials

- meterstick
- metric ruler
- spring scale
- triple-beam balance
- electronic scale (if available)
- 3 filled beakers labeled *1*, *2*, and *3*
- 3 filled beakers labeled *4*, *5*, and *6*
- 3 filled graduated cylinders labeled *7*, *8*, and *9*
- various classroom objects (listed in the "Data and Results" section)

Procedure: Read the entire procedure before you begin the investigation. Note that your teacher will prepare the labeled beakers and graduated cylinders for you to use. As you complete each part of the activity, record your data in the "Data and Results" section.

Name ______________________________ Date ______________

The Importance of Significant Figures (continued)

Part I: Measuring Length

1. Without using any measuring tools, estimate the lengths of the objects listed in Table A of the "Data and Results" section. Write your estimates in millimeters (mm) or centimeters (cm).
2. Measure each item by using the suggested measuring device, and record your values in the data table.
3. Determine the number of significant figures in each measurement by counting the doubtful digit (the last significant figure of the measurement) and all the other digits to the left of it, up to and including the last digit that is not zero. Record your results in the data table.

Part II: Measuring Mass

4. Without using any measuring tools, estimate the mass of the objects listed in Table B of the "Data and Results" section. Write your estimates in grams (gm).
5. Use a spring scale to measure the mass of each item, and record the values in the data table.
6. Measure each item again, this time with a triple-beam balance. Record the values in the data table.
7. If available, measure each item again with an electronic scale. Record the values in the data table.
8. Determine the number of significant figures in each measurement (as you did in Part I). Record the values in the data table.

Part III: Measuring Volume

9. Estimate the volume of liquid in beakers 1, 2, and 3. Write your estimates in milliliters (mL) in Table C of the "Data and Results" section. Note that you will not be measuring the actual volumes.
10. Measure the volume of liquid in beakers 4, 5, and 6 by reading the measurement mark (on the beaker) that aligns with the liquid's surface. Record your measurements in the data table. Note that you will not be making estimates.
11. Measure the volume of liquid in graduated cylinders 7, 8, and 9 (the same way you read the volumes in step 10). Record your measurements in the data table. Note that you will not be making estimates.
12. Record the number of significant figures in each measured value. Record your results in the data table.

Name ______________________________ Date ______________

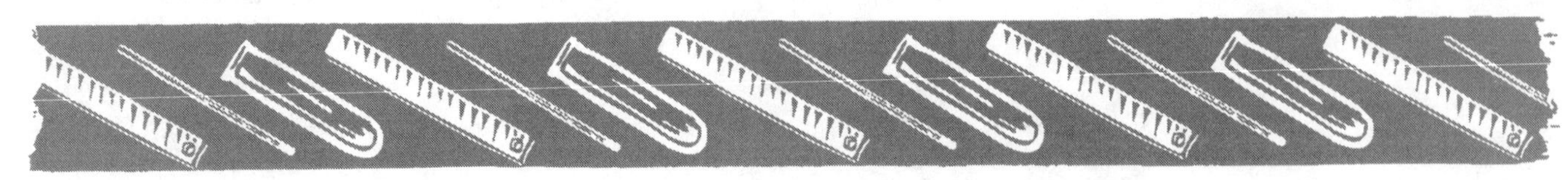

The Importance of Significant Figures (continued)

Data and Results

TABLE A

Item	Estimated Length	Measuring Device	Actual Length	# of Sig. Figures
paperback		metric ruler		
pencil		metric ruler		
your shoe		metric ruler		
chalkboard		meterstick		
room length		meterstick		
door length		meterstick		
desk height		metric ruler		

TABLE B

Item	Estimated Mass	Spring Scale	Balance	Electronic Scale	# of Sig. Figures
paperback					
pencil					
paper					
ruler					
scissors					
chalk					
glue bottle					

Name ______________________________ Date ____________

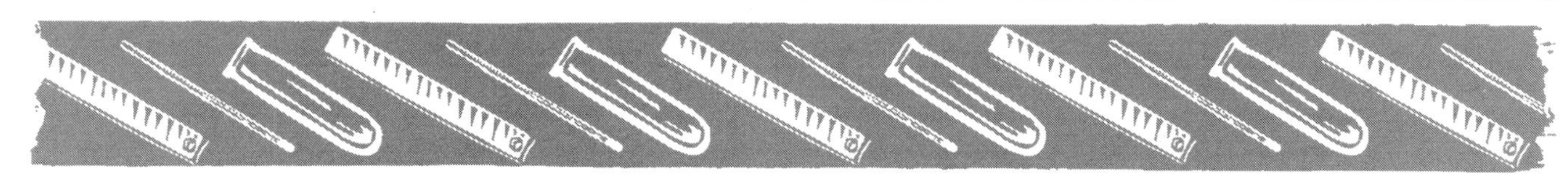

The Importance of Significant Figures (continued)

TABLE C

Kind of Container	Estimated Volume	Measured Volume	# of Sig. Figures
1			
2			
3			
4			
5			
6			
7			
8			
9			

Follow-Up

1. What instrument was more precise in measuring length? Why?

____

____

2. What instrument was more precise in measuring mass? Why?

____

____

3. What instrument was more precise in measuring volume? Why?

____

____

Name ______________________________ Date ______________

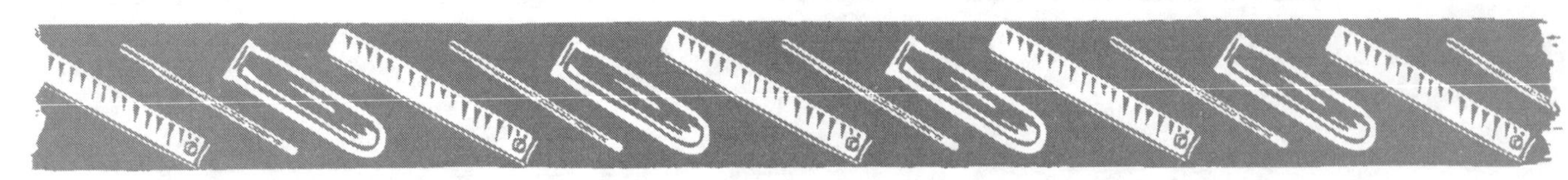

Devising Measuring Strategies

Purpose: Develop strategies to solve unique measuring problems.

Background: Sometimes items need to be measured, but direct measurements are either impossible or impractical due to the size or shape of the object. In these instances, indirect measurement techniques are used. For example, cargo ships have marks on the outside of their hulls that indicate how low or how high the ship is riding in the water, thus indirectly measuring the weight of the cargo inside the ship. This technique helps tugboat pilots to decide, at a glance, if a cargo ship can navigate a shallow harbor.

Materials

The materials you need will depend on your strategies (see the procedure).

Procedure: Work with one or two classmates. Read the entire procedure before you begin the investigation.

1. With your partners, decide how you will measure each item listed in Part I of the "Data and Results" section. Record your strategies.
2. In Part II, write a list of the materials your group will need to carry out the measurements. Give a copy of the list to your teacher, so he or she can help you gather the items. You may have to hold off on measuring some of the items until your teacher can obtain them, or you may need to revise and adapt your methods to suit the materials that are available.
3. Follow your strategies to complete the measurements. Record your values in the data table, Part III.
4. Compare your results with those of four other groups. Discuss the different strategies used. Then record in the data table the other groups' results, and calculate the average for each set of values.

Name ____________________ Date __________

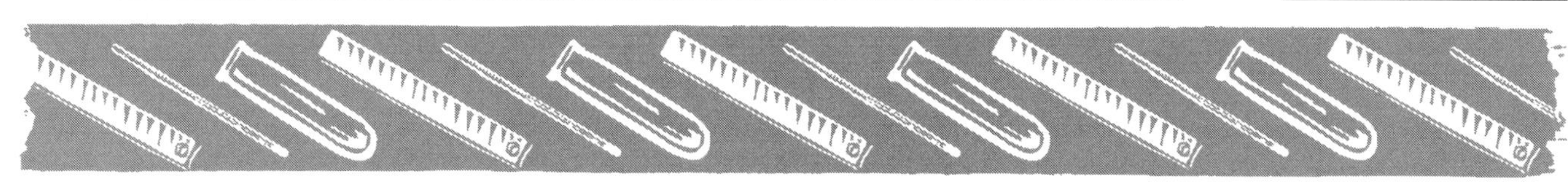

Devising Measuring Strategies (continued)

Data and Results

Part I. Measurement Strategies

The thickness of your hair.

Strategy: ____________________

The mass (weight) of a straight pin.

Strategy: ____________________

The area of the metal eraser holder on a pencil.

Strategy: ____________________

The mass (weight) of a grain of rice.

Strategy: ____________________

The volume of a single drop of water.

Strategy: ____________________

The thickness of a sheet of notebook paper.

Strategy: ____________________

Part II. Materials Needed

Name ______________________________ Date ______________

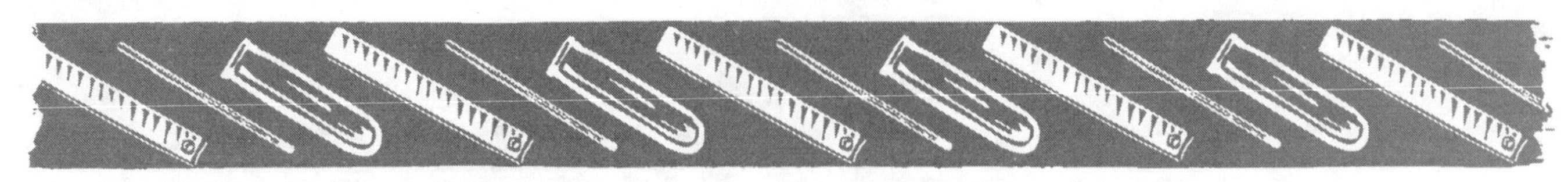

Devising Measuring Strategies (continued)

Part III. Measurements

Item	Your Results	Results from Other Groups 1	2	3	4	Average
hair						
straight pin						
eraser holder						
grain of rice						
drop of water						
sheet of paper						

Follow-Up

1. How did your strategies compare to those of your classmates? How were they similar? How were they different?

____

____

____

2. How did your results compare to those of your classmates'?

____

____

3. If you could do this investigation again, how would you modify your strategies?

____

____

____

Name ____________________ Date __________

Do You Overestimate or Underestimate Yourself?

Purpose: Make and check estimates that relate to measurements about yourself!

Background: A lot of extra time and work can be avoided by learning how to make good estimates. Estimates can help you determine the reasonableness of calculated answers. They can also be helpful in everyday life when numerical values are needed, but there's no time to calculate exact answers, like when you're standing in a checkout line of a grocery store, wondering if you have enough money to pay your bill. Just how good are you at making estimates, especially about yourself?

Materials

- meterstick
- string
- tape measure

Procedure: Read the entire procedure before you begin the activity.

1. Look at the items listed in the "Data and Results" section. Estimate each measurement about yourself, and then record your predictions.
2. Use the necessary tools to carefully measure the listed items. Record the values in the data table.
3. For each item, calculate the difference between the estimated value and actual measurement. Write the answer as a positive number (+) if the estimate was larger than the actual measurement; write the answer as a negative number (–) if the estimate was smaller.
4. After you complete the chart, add the positive and negative numbers to get the total difference. Record your answer under the data table. If the total is a negative number, you underestimated yourself; if it's a positive number, you overestimated yourself.

Name ______________________________ Date ______________

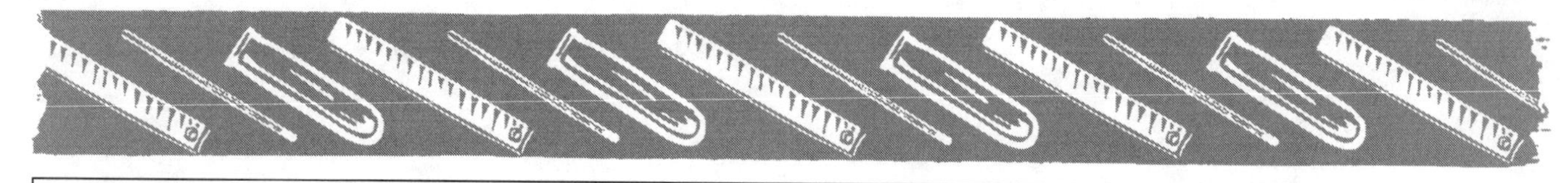

Do You Overestimate or Underestimate Yourself? (continued)

Data and Results

Feature to Measure	Estimate	Actual Size	Difference (+ or – number)
nose length			
smile			
height			
leg length			
foot length			
calf width			
waist			
arm length			
thumb length			
little finger circumference			
head circumference			
ear length			
neck circumference			
big toe circumference			

Total Difference: ____________

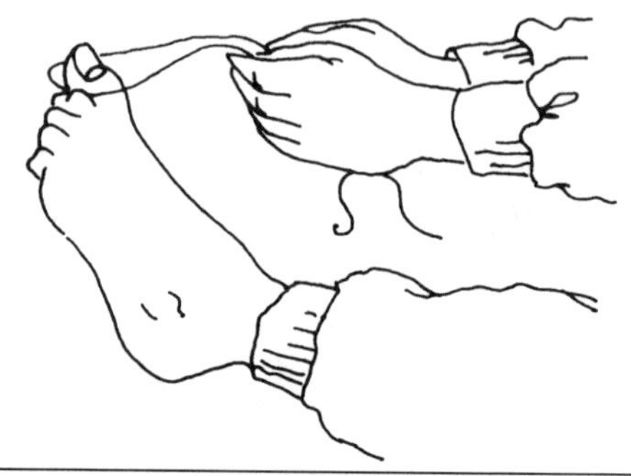

Name ______________________________ Date ______________

Determining the Moon's Diameter

Purpose: Construct an instrument to collect data about the moon, and use that information to calculate the moon's diameter.

Background: The moon is about 384,400 km from Earth, but it is possible to use a simple instrument to calculate the moon's diameter.

Materials

- meterstick
- 3" x 5" (7.5 cm x 12.5 cm) index card
- 5" x 7" (12.5 cm x 17.5 cm) index card
- X-acto® knife or single-edged razor blade
- pen
- masking tape
- aluminum foil
- straight pin
- flashlight

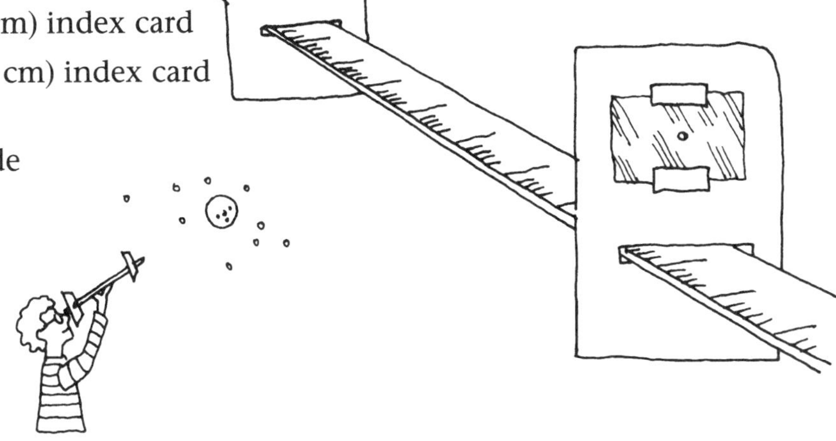

Procedure: Read the entire procedure before you begin. Refer to the above diagram to help you complete each step.

1. Place both the small and large index cards in a vertical position (long side upward). Mark 1 cm from the bottom of each card. Then at the 1-cm mark, cut a horizontal slit in each card so that the cards slide snugly but easily over the end of a meterstick. After checking to see if the slits are the appropriate size, remove the cards from the stick.
2. Using your meterstick, measure 4 cm from the top of the small card, and draw two parallel, vertical lines that are exactly 1 cm apart. The parallel lines should be centered above the slit.
3. Above the slit on the large index card, draw and cut out a "window" that is 5 cm tall and 8 cm wide. The window should be centered above the slit.
4. Tape aluminum foil over the window opening. Then use a pin to make a small hole in the center of the foil.
5. Insert the meterstick into the slit of the large card. Slide the card along the stick until it aligns with the 80-cm mark. Then slide the small card onto the other end of the stick at about the 20-cm mark.

Name ______________________________ Date ____________

Determining the Moon's Diameter (continued)

6. When there is a full moon, go outside to an area where there are no bright lights nearby and where you can clearly see the moon. Bring along your meterstick device, a flashlight, pencil, and sheet of paper.

   (Note: If there is a breeze outside, you may need to attach stiff cardboard or cardstock to the cards on the meterstick to keep them from bending.)
7. Aim the measuring device at the moon, with the large-card end closest to the moon. Align the measuring device so that the image of the moon comes through the pinhole and appears on the small card. Slide the small card back and forth until the image of the moon fits exactly between the two parallel lines. Make sure neither card is bending during this time.
8. Read the scale on the meterstick to determine the distance between the two cards. Record the distance rounded off to the nearest centimeter. (Note: Make sure to check where your small card is positioned. Don't assume that it's still at the 20-cm mark.)

Data and Results

1. When the measuring device was correctly aligned with the moon, what was the distance between the two cards?

   Distance between the two cards: ________cm
2. Use your measured value and the following formula to calculate the diameter of the moon. Note that the distance from Earth to the moon is approximately 384,400 km.

$$\frac{\textbf{Diameter of the moon's image (1.0 cm)}}{\textbf{Distance between the cards (cm)}} = \frac{\textbf{Diameter of the moon (km)}}{\textbf{Distance to the moon (km)}}$$

   Calculated diameter of the moon: ________________
3. Find out the actual diameter of the moon from your teacher. Compare the actual diameter to your calculated value. How accurate was your measurement?

   ____
4. What possible errors could you have made in your investigation? How might you improve your results?

   ____

   ____

Name ______________________________ Date ______________

Scale Models of Planets and Moons

Purpose: Calculate the diameter of celestial bodies by using scale models.

Background: The term *scale* refers to a ratio (a:b or a/b) that compares the dimensions of a model, drawing, or map to those of the object being represented. For example, a model car designed at a 1:48 scale is one in which 1 unit (such as 1 inch) of the model represents 48 units of the real car. The ratio of the scale depends upon the size of the object being represented. For example, it wouldn't be practical to use a 1:2 scale for a map of New Mexico.

Materials

- meterstick
- pencil
- 4-m strip of adding-machine tape

Procedure: Work with a partner. Read the entire procedure before you begin the activity.

1. Complete Table A of the "Data and Results" section by calculating the scale diameter (in centimeters) of the various celestial bodies. To calculate the values, use the scale 1 cm:100 mi. For example, a 200-mi. diameter would be represented by a scale diameter of 2 cm.
2. Lay a strip of adding-machine tape flat on the floor or on a large table. Draw a line across one end of the paper strip, close to the edge. Label the line *Earth Edge 1.*
3. From the line, use a meterstick and a pencil to measure and mark off on the paper strip the length of the scale diameter of Earth (listed in Table A). Label this second line *Earth Edge 2.* The distance between the two lines represents the scale diameter of Earth.
4. Measure again from the *Earth Edge 1* line, but this time mark off the scale diameter of the earth's moon. Label this line *Moon Edge.* The distance between *Earth Edge 1* and *Moon Edge* represents the scale diameter of the moon. Notice the ease with which you can compare the relative sizes of Earth and its moon, just by looking at the scale model.
5. Turn the paper strip over to the other side, and follow the same procedure to construct another scale model, one that compares the size of Uranus to its moons. (Mark off and label *Uranus Edge 1, Uranus Edge 2, Miranda Edge,* and so on, to complete the model.)

Name ______________________________ Date ______________

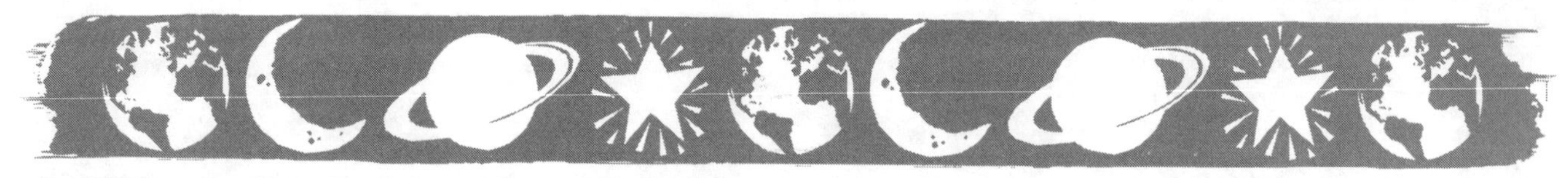

Scale Models of Planets and Moons (continued)

Data and Results

TABLE A

(Scale: 1 cm = 100 miles)

Celestial Body	Actual Diameter (rounded off)	Scale Diameter
Earth	7,900 miles	
Earth's Moon	2,200 miles	
Uranus	32,000 miles	
Uranus's Moons		
Miranda	200 miles	
Ariel	500 miles	
Umbriel	350 miles	
Titania	600 miles	
Oberon	500 miles	

Follow-Up

1. Why was adding-machine tape used to construct these scale models?

____

____

2. How could you use these models to determine how much larger or smaller one celestial body is compared to another?

____

____

3. If the scale used in this activity was 1 cm:1,000 mi., how would that affect the scale sizes of the planets and moons?

____

____

Name ______________________ Date __________

Scale Models of Planets and Moons (continued)

4. Why is it important to choose a particular scale when making a model of an object?

____

____

5. When the scale of a model is 1:1, what does that mean?

____

____

6. Why would a model of a sailing ship be 1:256, but a model of a car be 1:48?

____

____

7. What would a model airplane look like if its parts were not all made to the same scale?

____

____

8. **Technology Connection:** Many computer programs today (especially those that have CAD ability—computer-assisted drawing ability) enable users to construct or view three-dimensional scale models of virtually any object. Investigate how computer designers and movie producers use these programs to create interactive games and special effects. If possible, examine computer programs that generate scale models. Write a summary about what you learned.

____

____

____

____

____

____

Name ______________________________ Date ______________

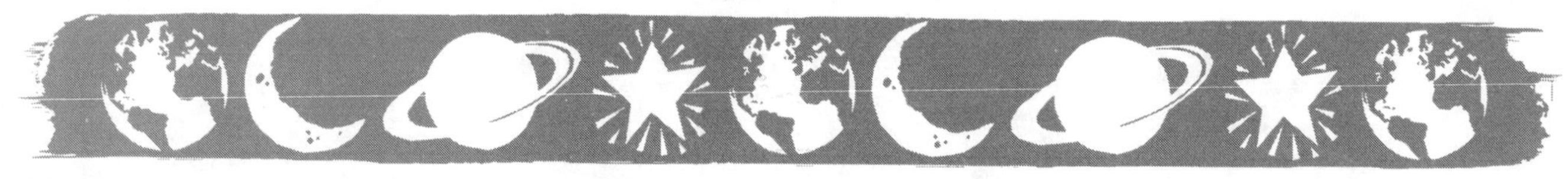

Scale Models of Planets and Moons (continued)

Extension Activity

Complete the following table. Then use a 4-m strip of adding-machine tape, a meterstick, and a pencil to make two scale models—one representing the relative distance of each planet from the sun, and the other representing the relative diameters of the planets. When you are done, use your scale models and a meterstick (not the data table) to answer the questions at the bottom of the page.

Planet	Distance from the Sun (miles)	Scale Distance (1 cm:10 million mi.)	Actual Diameter (miles)	Scale Diameter (1 cm: 250 mi.)
Mercury	36,000,000		3,000	
Venus	67,000,000		7,500	
Earth	93,000,000		7,900	
Mars	142,000,000		4,200	
Jupiter	483,000,000		89,000	
Saturn	886,000,000		76,000	
Uranus	1,782,000,000		32,000	
Neptune	2,793,000,000		28,000	
Pluto	3,670,000,000		1,450	

1. Which planet is farthest from the sun? How far is it from the sun?

____

2. Which planet is closest to Earth? How many million miles is it from Earth?

____

3. How many million miles is it from Neptune to Pluto? from Mars to Jupiter?

____

4. Which planet is closest to Earth's size? What is the difference in their diameters?

____

5. How many miles larger is Jupiter than Saturn? Venus than Mars?______________

6. What are the names of the planets, in order from smallest to largest?

____

Name ______________________________ Date ______________

Star Light, Star Bright

Purpose: Compare the brightness and surface temperatures of various stars by plotting their values on a semi-logarithmic graph.

Background: The hottest stars in the galaxy are blue-white in color, whereas the stars with the lowest temperature appear red. Other stars fall within these two extremes, ranging in surface temperature from 1,500°C to over 40,000°C. The actual brightness of a star—its absolute magnitude—is based on a comparison with the sun's illumination, which is assigned a brightness value of 1. A star 10 times as bright as the sun has an absolute magnitude of 10. The Hertzsprung-Russell Diagram is a graph on which the surface temperatures of stars are plotted against their luminosities (brightness).

Materials

- Star Chart
- Hertzsprung-Russell Diagram
- colored pencils
- metric ruler

Procedure: Read the entire procedure before you begin the activity.

1. Use the data from the Star Chart to plot three stars on the Hertzsprung-Russell Diagram. Draw a dot for each star, and label the point by using the first three letters of the star's name. (Note: For the Sirius and Ross stars, also include the letter or number of the star.)
2. To make sure that you are reading the temperature and brightness scales correctly, have your teacher confirm the placement of the three stars you plotted on the graph. Then plot the values of the remaining stars. Note that some of your plotted points will be approximations, since some of the stars' values fall between the listed values on the graph.
3. Darken your plot points with a pencil or pen. Then use colored pencils to shade portions of the graph to match the color of the stars that fall within that temperature range: light red, 2,000°C to 3,500°C; orange-red, 3,500°C to 5,000°C; yellow-white, 5,000°C to 6,000°C; blue-white, 6,000°C to 7,500°C; and blue, 7,500°C to 40,000°C.

Name ______________________________ Date ______________

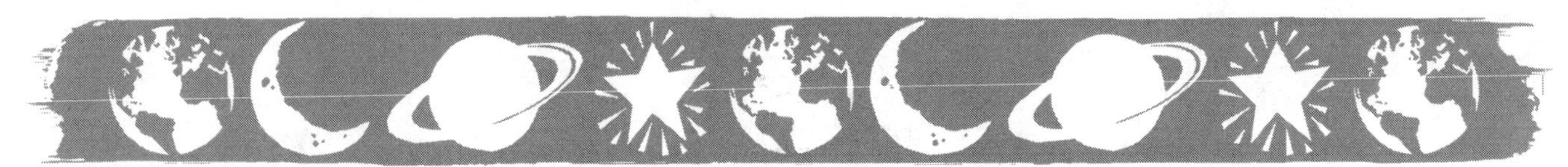

Star Light, Star Bright (continued)

Data and Results

1. What happens to the temperature as you move to the right on your graph?

   ____

2. What happens to the brightness as you move from top to bottom on your graph?

   ____

3. How would you describe a star in the upper left-hand corner of your graph?

   ____

   ____

Follow-Up

1. Stars are classified in spectral classes, according to their temperatures. Use the following information (approximate values) to label the spectral classes on your graph.

Class of Star:	O	B	A	F	G	K	M
Temp. (°C x 1,000):	30+	12–30	8–12	6–8	4.5–6.0	3.5–4.5	2.5–3.5

2. Some stars are known as "red giants" or "white dwarfs" because of their appearance. Red giants can be 100 times bigger than our sun, but they are much cooler. White dwarfs, on the other hand, are about the same size as Earth, but they are much hotter than our sun. Include this information on your graph by completing the following steps:

   a. Look for the red giants in the red area of the graph. Lightly circle and label the group of stars.

   b. Look for the white dwarfs in the blue-white area of the graph. Lightly circle and label the group of stars.

   c. Circle and label the remaining stars, called *main sequence stars*, which fall in a diagonal pattern from the upper left to the lower right of the graph.

Name ______________________ Date __________

Star Light, Star Bright (continued)

STAR CHART

Name of Star	Surface Temp. (°C)	Brightness (absolute magnitude)
Beta Crucis	35,000	5,000
Spica	22,000	1,000
Rigel	12,000	40,000
Vega	13,000	80
Sirius A	11,000	23
Sirius B	13,000	0.008
Deneb	9,000	30,000
Canopus	7,500	1,000
Altair	7,300	10
Capella	5,400	200
Aldeberan	4,000	80
Barnard's Star	2,800	0.00044
Betelgeuse	2,800	20,000
Ross 154	2,800	0.004
Ross 248	2,700	0.00011
Epsilon Eridani	4,500	0.30
Ross 128	2,800	0.00033
Proxima Centauri	5,800	1.3
Sun	5,600	1
Antares	3,200	4,000

Name ______________________ Date ____________

Star Light, Star Bright (continued)

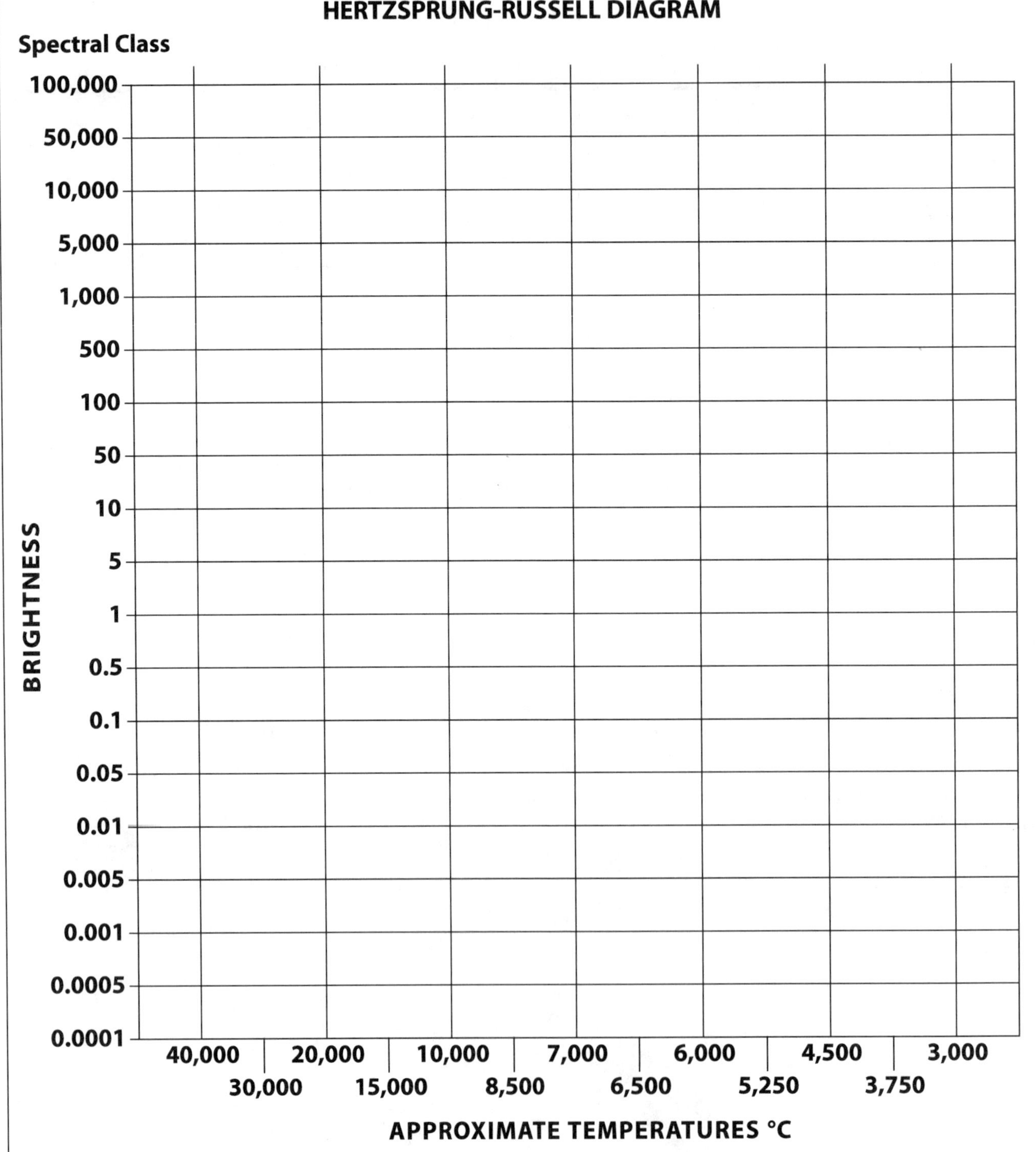

Name ______________________________ Date ______________

Predicting Sunspots

Purpose: Graph data about sunspots, and look for patterns or cycles in the plotted values to make predictions about future sunspot activity.

Background: Sunspots are relatively cool, dark areas of the sun's surface. Usually beginning as small dots on the surface of the sun, they may quickly expand to thousands of kilometers in size. Sunspots can last anywhere from a day up to several months.

Materials

- Sunspots Table
- graph paper
- calculator

Procedure: Read the entire procedure before you begin the investigation.

1. Plot the values listed on the Sunspots Table. Set up the graph to show "year" on the horizontal axis and "number of sunspots" on the vertical axis. Connect the plotted points to make a line graph.
2. Use the information from your graph to complete the "Data and Results" section. Note that a peak in the line graph represents a year of maximum sunspot activity, and the lowest point of a valley represents a year of minimum activity.

Data and Results

1. What is the average number of years between sunspot maximums?

____

2. What is the average number of years between sunspot minimums?

____

3. What is the greatest number of years between consecutive sunspot maximums?

____

4. What is the fewest number of years between consecutive sunspot maximums?

____

Name ______________________________ Date ______________

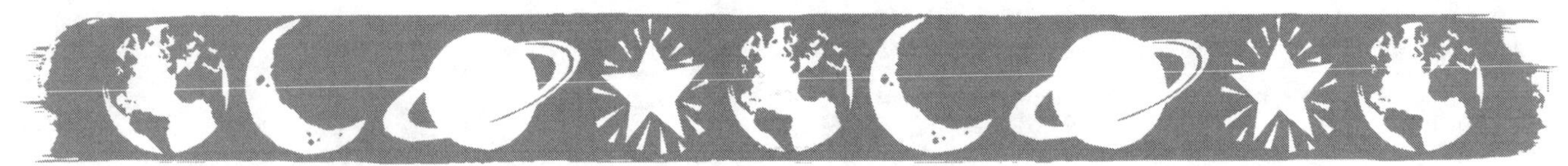

Predicting Sunspots (continued)

5. What is the greatest number of years between consecutive sunspot minimums?

____

6. What is the fewest number of years between consecutive sunspot minimums?

____

7. Based on the data, when do you think the next solar maximum occurred after 1984? When do you think the next solar minimum occurred?

____

____

8. When was the last solar maximum before 1884?

____

9. When was the last solar minimum before 1884?

____

10. How many sunspots were there the year you were born? Estimate the number.

____

11. Sunspot activity has been known to disrupt radio communications. World War II was the first global conflict in which reliable radio communication was essential to the armed forces of all countries involved. Do you think sunspot activity during this time was particularly high? What effect do you think it had upon that conflict?

____

____

____

____

____

Name ______________________________ Date ____________

Predicting Sunspots (continued)

SUNSPOTS TABLE

Year	# of Sunspots	Year	# of Sunspots	Year	# of Sunspots
1884	64	1920	37	1956	142
1886	25	1922	14	1958	185
1888	6	1924	16	1960	112
1890	7	1926	64	1962	38
1892	72	1928	78	1964	10
1894	78	1930	36	1966	47
1896	41	1932	11	1968	106
1898	26	1934	9	1970	105
1900	9	1936	80	1972	69
1902	5	1938	110	1974	35
1904	42	1940	68	1976	13
1906	53	1942	31	1978	92
1908	48	1944	101	1980	155
1910	18	1946	93	1982	116
1912	3	1948	136	1984	46
1914	9	1950	84		
1916	57	1952	32		
1918	80	1954	4		

Name ______________________________ Date ______________

How Much Humidity?

Purpose: Measure the relative humidity of the air.

Background: The term *humidity* describes the amount of moisture in the air. *Relative humidity* (expressed as a percentage) is a measurement that compares the actual amount of moisture in the air to the maximum amount of moisture the air can absorb at that temperature. Warm air can hold more moisture than cold air, which means that 50% humidity at 30°C represents more moisture in the air than 50% humidity at 15°C. When a mass of air is cooled, its relative humidity increases (because the maximum amount of moisture the air can hold decreases). At the point when the relative humidity reaches 100%, the air mass is saturated and water droplets will form in the air.

Materials

- 2 Celsius thermometers
- white, wide, hollow shoestring
- white thread
- scissors
- small beaker of water
- Relative-Humidity Chart

Procedure: Work with a partner. Read the entire procedure before you begin the investigation.

1. Cut off a piece of shoestring that matches the length of a thermometer bulb, from the bottom of the bulb to the 0-degree mark. Slip the hollow piece of shoestring over the bulb of one of your thermometers (like you would a sock), and tie string around the ends of the covering, above and below the bulb, to keep the covering from slipping off. This thermometer will be your "wet-bulb thermometer." The second, uncovered thermometer will be your "dry-bulb thermometer."
2. Take both thermometers outside, and use a beaker of water to wet the covered end of the wet-bulb thermometer. Place both thermometers in the same location and wait until the temperature of the wet-bulb thermometer stops falling (it should take about three minutes). Then read the temperature off both thermometers, and record the values in the "Data and Results" section.

Name ____________________ Date ____________

How Much Humidity? (continued)

3. Calculate the difference between the two temperature readings, and record the value in the "Data and Results" section. Then determine the relative humidity of the air by using the Relative-Humidity Chart as follows:

   a. On the left side of the chart, locate the value (from 0 to 30) that matches the dry-bulb reading.

   b. At the top of the chart, locate the value (from 1 to 17) that matches the temperature difference between the dry-bulb and the wet-bulb readings.

   c. Point to both values on the chart at the same time (with different hands), and slide your fingers across or down the lists of numbers to find the point of intersection. The number at that point of intersection is the relative humidity of the outside air. For example, a dry temperature of 8°C and a temperature difference of 6°C yields a relative humidity of 29%.

Data and Results

Temperature of the dry-bulb thermometer: ________

Temperature of the wet-bulb thermometer: ________

Difference in temperature between the dry-bulb and wet-bulb readings: ________

1. Based on your results, what is today's relative humidity?

   ____

2. If the temperature outside went up 10 degrees and nothing else changed, what would happen to the relative humidity?

   ____

3. What is the relative humidity when the wet- and dry-bulb temperatures are the same?

   ____

4. Why isn't relative humidity measured according to how many milliliters of water are in a cubic meter of air?

   ____

Name ______________________________ Date ______________

How Much Humidity? (continued)

RELATIVE-HUMIDITY CHART

Dry-Bulb Temp. ↓ / **Difference in Temp. between Wet-Bulb and Dry-Bulb Readings**

°C	1	2	3	4	5	6	7	8	9	10	11	12	13	14	15	16	17	18
0	**81**	**64**	**46**	**29**	**13**													
1	**83**	**66**	**49**	**33**	**17**													
2	**84**	**68**	**52**	**37**	**22**	**7**												
3	**84**	**70**	**55**	**40**	**26**	**12**												
4	**85**	**71**	**57**	**43**	**29**	**16**												
5	**86**	**72**	**58**	**45**	**33**	**20**	**7**											
6	**86**	**73**	**60**	**48**	**35**	**24**	**11**											
7	**87**	**74**	**62**	**50**	**38**	**26**	**15**											
8	**87**	**75**	**63**	**51**	**40**	**29**	**19**	**8**										
9	**88**	**76**	**64**	**53**	**42**	**32**	**22**	**12**										
10	**88**	**77**	**66**	**55**	**44**	**34**	**24**	**15**	**6**									
11	**89**	**78**	**67**	**56**	**46**	**36**	**27**	**18**	**9**									
12	**89**	**78**	**68**	**58**	**48**	**39**	**29**	**21**	**12**									
13	**89**	**79**	**69**	**59**	**50**	**41**	**32**	**23**	**15**	**7**								
14	**90**	**79**	**70**	**60**	**51**	**42**	**34**	**26**	**18**	**10**								
15	**90**	**80**	**71**	**61**	**53**	**44**	**36**	**27**	**20**	**13**	**6**							
16	**90**	**81**	**71**	**63**	**54**	**46**	**38**	**30**	**23**	**15**	**8**							
17	**90**	**81**	**72**	**64**	**55**	**47**	**40**	**32**	**25**	**18**	**11**							
18	**91**	**82**	**73**	**65**	**57**	**49**	**41**	**34**	**27**	**20**	**14**	**7**						
19	**91**	**82**	**74**	**65**	**58**	**50**	**43**	**36**	**29**	**22**	**16**	**10**						
20	**91**	**83**	**74**	**66**	**59**	**51**	**44**	**37**	**31**	**24**	**18**	**12**	**6**					
21	**91**	**83**	**75**	**67**	**60**	**53**	**46**	**39**	**32**	**26**	**20**	**14**	**9**					
22	**92**	**83**	**76**	**68**	**61**	**54**	**47**	**40**	**34**	**28**	**22**	**17**	**11**	**6**				
23	**92**	**84**	**76**	**69**	**62**	**55**	**48**	**42**	**36**	**30**	**24**	**19**	**13**	**8**				
24	**92**	**84**	**77**	**69**	**62**	**56**	**49**	**43**	**37**	**31**	**26**	**20**	**15**	**10**	**5**			
25	**92**	**84**	**77**	**70**	**63**	**57**	**50**	**44**	**39**	**33**	**28**	**22**	**17**	**12**	**8**			
26	**92**	**85**	**78**	**71**	**64**	**58**	**51**	**46**	**40**	**34**	**29**	**24**	**19**	**14**	**10**	**5**		
27	**92**	**85**	**78**	**71**	**65**	**58**	**52**	**47**	**41**	**36**	**31**	**26**	**21**	**16**	**12**	**7**		
28	**93**	**85**	**78**	**72**	**65**	**59**	**53**	**48**	**42**	**37**	**32**	**27**	**22**	**18**	**13**	**9**	**5**	
29	**93**	**86**	**79**	**72**	**66**	**60**	**54**	**49**	**43**	**38**	**33**	**28**	**24**	**19**	**15**	**11**	**7**	
30	**93**	**86**	**79**	**73**	**67**	**61**	**55**	**50**	**44**	**39**	**35**	**30**	**25**	**21**	**17**	**13**	**9**	**5**

Name ______________________________ Date ______________

Weather Watch

Purpose: Predict future weather conditions by collecting weather data and analyzing the changes in temperature, barometric pressure, relative humidity, wind speed, and wind direction.

Background: Weather is affected by a variety of atmospheric conditions such as temperature, humidity, and wind. Meteorologists use a variety of tools to analyze the changes in these conditions and to predict the weather. With the help of sophisticated computers and other modern technology, meteorologists are able to study the physics of storms, the patterns of jet streams, and the consequences of natural phenomenon on global weather.

Materials

- several copies of the Weather Watch Chart
- wet-bulb and dry-bulb thermometers
- Relative-Humidity Chart
- standard aneroid barometer
- wind meter
- weather vane
- photographs of cloud types (or reference books that provide this information)

Procedure: Work with two or three classmates. Read the procedure before you begin the activity. If needed, review the activity *How Much Humidity?* to recall how to make wet-bulb and dry-bulb thermometers.

1. Prepare for the investigation: make the wet- and dry-bulb thermometers, learn how to use the weather tools, review how to determine relative humidity, and study pictures of clouds until you can easily identify the different types (such as cirrus, stratus, cumulus, nimbus, and cumulonimbus) by sight. Check with your teacher about the time frame for this activity.
2. Decide who will be responsible for taking the different measurements each day. Rotate responsibilities throughout the investigation. Always measure the weather at the same time of day.
3. Use photocopies of the Weather Watch chart to record the data your group collects. Look for patterns or connections between measured data and outside weather conditions. Refer to the following guidelines as you complete your weather chart.

Name ______________________________ Date ______________

Weather Watch (continued)

Guidelines for Completing the Weather Watch Chart

- If you are using Fahrenheit thermometers, remember to convert the measurements to Celsius before you record the values on the chart. The formula is °C = (°F – 32)/1.8.
- Use the Relative-Humidity Chart to help you determine the relative humidity. If you do not remember how to calculate relative humidity, review the activity *How Much Humidity?* on page 56.
- Read the barometer as explained by your teacher, and express the answer in inches of mercury.
- Use the wind meter to determine wind speed. Use the weather vane to determine wind direction. Remember that the wind direction is the direction the wind is blowing from. A feather blown by a northerly wind flies south.
- Classify the clouds as *cirrus, cumulus,* or *stratus* based on their shape and appearance.
- Estimate the percentage of the sky covered by clouds.
- Describe the weather conditions in general terms such as *raining, windy, stormy, freezing cold, fair,* and *hot.*

Data and Results

1. What type of weather is associated with low barometric pressure? with high barometric pressure?

   ____

2. What other patterns do you notice between the recorded measurements and the general weather conditions each day?

   ____

3. What one piece of weather information is the most important to gather?

   ____

4. Describe the environment within 20 miles of your school. Pay attention to the presence of large, plant-covered areas, hills, and bodies of water. How do you think this surface environment affects the weather? If your school is in a metropolitan area, how would the weather be different if the land was in its original, natural state?

Name ______________________________ Date ______________

Weather Watch (continued)

WEATHER WATCH CHART

	1	2	3	4	5
Date					
Time of day					
Dry-bulb temperature					
Wet-bulb temperature					
Relative humidity					
Barometric pressure					
Wind speed					
Wind direction					
Cloud type					
Cloud cover (%)					
Description of weather conditions					

Name ______________________________ Date ____________

Survival of the Fittest

Purpose: Demonstrate how the size of an animal may be a factor in natural selection.

Background: The process of natural selection was first suggested by Charles Darwin to explain why some species survive and others become extinct. The theory states that the organisms best adapted to their environment are the ones that survive.

Materials

- colored construction paper (choose only one color)
- metric ruler
- scissors
- large container
- calculator

Procedure: Work with three classmates. Read the entire procedure before you begin the activity.

1. Cut out three sets of 50 squares: 2-cm squares, 4-cm squares, and 8-cm squares. Each group of squares should be uniform in size and shape. These 150 squares will represent populations of small, medium, and large rodents.
2. Place the 150 "rodents" into a container, and mix them up thoroughly without bending them. Then randomly scatter them onto a table, which will represent the rodents' environment.
3. You and your partners are considered these rodents' "predators." At the same time, each group member should "prey" upon the rodents, picking them up one at a time. Each person should stop when he or she has gathered five of them. (Note: The first person to touch a rodent gets it.)
4. Have each group member sort his or her rodents by size (small, medium, and large), and then record the number of each kind on Table I of the "Data and Results" section.
5. Determine how many of each-sized rodent is left on the table, and record the values in Table II. Then calculate the values as a percentage of the total (150), and record that data as well.
6. Compare your results with those of other groups in the class. Add up the total number of each-sized rodent that survived in the entire class, and record the information in Table III. Then calculate the values as a percentage of the total (150 x number of student groups), and record that data as well.

Name ______________________________ Date ______________

Survival of the Fittest (continued)

Data and Results

TABLE I

Student names:	________	________	________	________
# Small prey:	________	________	________	________
# Medium prey:	________	________	________	________
# Large prey:	________	________	________	________

TABLE II

	Small	**Medium**	**Large**
Number of rodents remaining:	________	________	________
Percentage of the total (150):	________	________	________

TABLE III

	Small	**Medium**	**Large**
Number of rodents remaining in the class:	________	________	________
Percentage of the total (150 x # groups):	________	________	________

Name ______________________________ Date ______________

Survival of the Fittest (continued)

Follow-Up

1. How did your group's results compare to those of the class? Which population of rodents has the greatest amount of survivors?

____

2. Why do you think the predation rates vary for different-sized rodents?

____

____

3. How could the size of an animal affect its chances for survival? Give a real-life example to support your answer.

____

____

4. Based on your results, which group of rodents is likely to produce the greatest number of offspring in the next generation, assuming litter size is not a variable? Explain your answer.

____

____

5. How do you think camouflage plays a role in natural selection? How could you test your theory?

____

____

____

6. What traits, other than size, do you think are important in natural selection?

____

____

Name ______________________________ Date ______________

A Salty Solution

Purpose: Determine how the mass and volume of salt and water are affected when the two substances are combined.

Background: The ocean is a great mass of salt water. Dissolving salt in water results in a physical change, but not a chemical one. One taste of the mixture tells you that salt is still present, although you can't see the crystals anymore. But what happens to the volumes and masses of the two substances that are combined?

Materials

- 100-mL graduated cylinder
- triple-beam balance or electronic scale
- 150-mL beaker
- water
- salt

Procedure: Work with two or three classmates. Read the entire procedure before you begin the investigation.

1. Determine the mass of an empty graduated cylinder, and then record the measurement in the "Data and Results" section.
2. Add exactly 30 mL of salt to the graduated cylinder. Measure the mass of the graduated cylinder filled with salt. Subtract the mass of the empty cylinder from the total to determine the mass of 30 mL of salt. Record both the mass and the volume of the salt. Then pour the salt into a beaker, and put it aside for later use.
3. Pour exactly 70 mL of water into the graduated cylinder. Measure the mass of the graduated cylinder filled with water. Then subtract the mass of the empty cylinder from the total to determine the mass of 70 mL of water. Record both the mass and the volume of the water.
4. Add the water to the salt in the beaker, and gently swirl the beaker until all the salt is dissolved. Measure the volume of the saltwater solution by pouring it back into the graduated cylinder. Record the volume of the solution.
5. Calculate and record the mass of the solution by following the same procedure you used for the others.

Name ______________________________ Date ______________

A Salty Solution (continued)

Data and Results

Mass of the graduated cylinder: __________

Mass of the salt: __________ Volume of the salt: __________

Mass of the water: __________ Volume of the water: __________

Mass of the solution: __________ Volume of the saltwater solution: __________

1. Add the salt's mass and the water's mass together. What is the sum? __________
2. Compare the calculated sum of the masses to the measured mass of the solution. Is there a difference?

   ____

3. Add the salt's volume and the water's volume together. What is the sum?__________
4. Compare the sum of the salt and water volumes to the solution's measured volume. What is the difference, if any?

   ____

5. Explain what happens to volume and mass when a substance is dissolved.

   ____

   ____

   ____

6. Would the same thing happen if alcohol was mixed with water? Why?

   ____

   ____

   ____

   ____

Name ______________________________ Date ______________

The Parallax Principle

Purpose: Use the principle of parallax to measure the distance from a standing position to a drawn line.

Background: When an object, such as a star or a distant mountain, is viewed from different places, its position seems to shift. This apparent shift is called *parallax*. You can easily observe parallax by performing this simple test: Close your right eye and point to an object across the room; your arm should be extended in front of you at eye level. Without moving your arm or finger, open your right eye and close your left. Notice that the object seems to have moved to the right!

Materials

- chalkboard or butcher paper
- chalk or markers
- metric ruler
- meterstick
- masking tape
- calculator

Procedure: Work with a partner. Read the entire procedure before you begin this investigation. Note that your teacher will decide whether you should use a chalkboard or posted butcher paper for this activity.

1. Draw at eye level a long, straight line across the chalkboard or posted butcher paper. Then stand in front of the line, a good distance away (beside your desk or at a spot designated by your teacher), while your partner stands next to it.
2. Hold one end of the ruler at the bridge of your nose, between your eyes. (You'll look a bit like a unicorn, but don't worry about it. Just don't poke the ruler in your eye.) Rest the other end of the ruler on top of the knuckles of your other hand, and extend your index finger to make a "pointer."
3. Close your right eye, and aim your fingertip toward the right end of the line. Without opening your eye or moving your finger, direct your partner to the place on the line where your finger is pointing, and have him or her mark the spot with an *X*.

Name ________________________________ Date ____________

The Parallax Principle (continued)

4. Without moving your finger, head, or body, open your right eye and close your left. Your finger has now apparently moved, and is pointing toward another place on the line. Direct your partner to that location on the line, and ask him or her to write a *Y* at that spot.
5. Tell your partner to use tape to mark the floor underneath your extended fingertip. Then walk to the line, and measure the distance between points X and Y. Write the value above the line.
6. Complete the "Data and Results" section to determine the distance from where you stood to the drawn line. Then switch roles with your partner, and repeat the activity.

Data and Results

Look at the diagram below, which shows the setup of this investigation. Note how your eyes and the tip of your finger made a triangle. Also note that a similar triangle (same shape, different size) was formed between the tip of your finger and the drawn line. Keep in mind that the parts of similar triangles are proportional to each other.

Length *b* is the distance between your eyes. Length *d* is the distance between points *X* and *Y* on the drawn line. For this example of parallax, the most important dimension to note is the altitude of each triangle. The altitude of the smaller triangle (length *a*) is the distance from the bridge of your nose to the tip of your finger. The altitude of the larger triangle (length *c*) is the distance from your fingertip to the line, which is the value you are trying to determine.

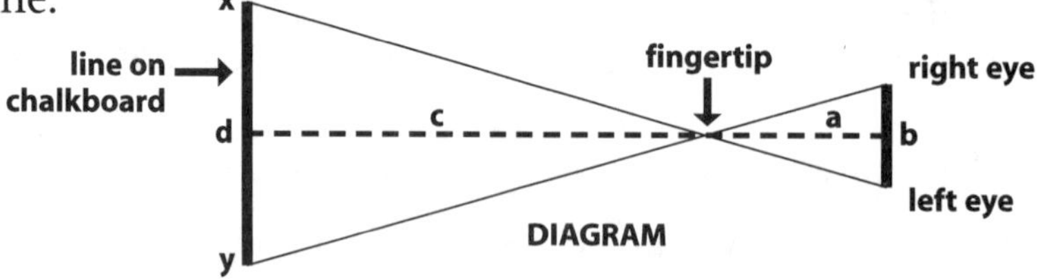

1. Distance between points X and Y (the measurement you wrote by the drawn line): ____________

2. Distance from the bridge of your nose to the tip of your finger (add the length of the ruler to the length of your index finger, from the knuckle to the tip): ____________

3. Distance between the pupils of your eyes: ____________

   (Write the value of 6 cm for the distance between your eyes, or have your partner carefully measure the distance between the pupils of your eyes.)

Name ______________________________ Date ______________

The Parallax Principle (continued)

4. The ratio of the altitude (length a) and base (length b) of the smaller triangle equals the ratio between the altitude (length c) and base (length d) of the larger triangle:

   $a/b = c/d$

   Determine the distance from your fingertip to the drawn line by plugging the answers from problems 1–3 into the equation and solving for c.

   Calculated distance from your fingertip to the line: _______

5. Check your answer by using a meterstick to measure the distance from the tape mark to the drawn line. Compare the calculated distance to the measured (actual) distance.

   Measured distance from your fingertip to the line: _______

   Difference between the calculated value and the actual value: _______

Follow-Up

1. How did your results compare to your partner's results?

   ____

2. What factors could have caused errors in your measurements?

   ____

   ____

3. In the space below, design an instrument for measuring parallax. (Hint: Think about the shape it could be.)

Name ________________________________ Date ______________

Angular Measurements

Purpose: Measure angles by using a protractor and a magnetic compass.

Background: Maps must be carefully drawn to scale to accurately show the distances being represented. But you also need to indicate directions with precision to make your map usable. A change in direction can be recorded by calculating the angular measurement—the amount of arc formed if a circle is drawn around the point where the direction changes. (See Example 1). The basic unit of angular measurement is the degree (°), which can be measured by a magnetic compass or a protractor. A complete circle is made up of 360 degrees.

Materials

- protractor
- magnetic compass
- metric ruler

Procedure: Read the entire procedure before you begin the activity.

1. Look at figures A–F in the "Data and Results" section. Use a protractor to measure the outer and inner angles formed in each figure (indicated by the arrows). Check to be sure that the protractor is properly aligned with the angle's point of intersection. You may choose to extend the lines of the angle to measure the figure more easily. (See Example 2). Write your measurements next to the corresponding angles.
2. Look at figures G–L in the "Data and Results" section. Use a protractor to measure and construct an angle equal to the number of degrees written for each figure. (See Example 3).
3. Look at figures M–O in the "Data and Results" section. The three circles represent magnetic compasses. On each circle, label *S*, *E*, and *W*, and add a compass needle pointing north. Then draw a second arrow to show the compass heading that is written below each figure. Note: If you orient a magnetic compass so the appropriate needle is pointing north, and then start walking north, you are following a compass heading of 0° (or 360°); a heading of 45° would take you in a northeasterly direction.

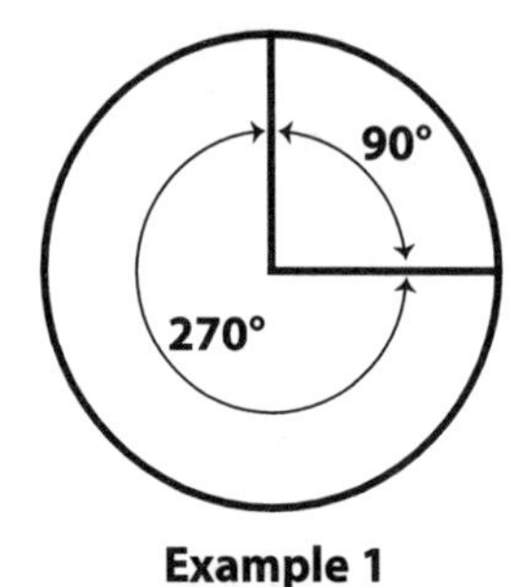

Example 1

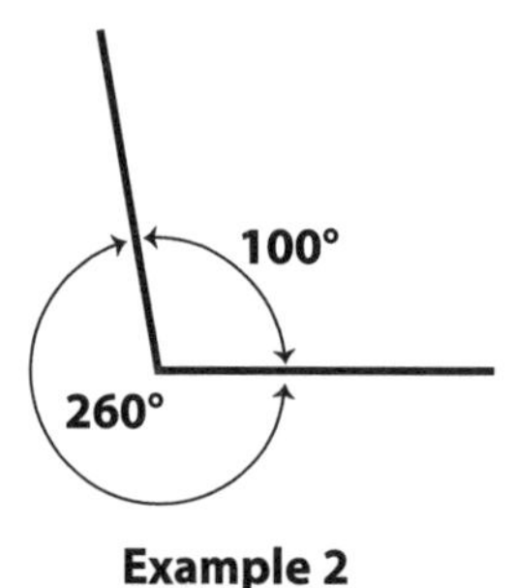

Example 2

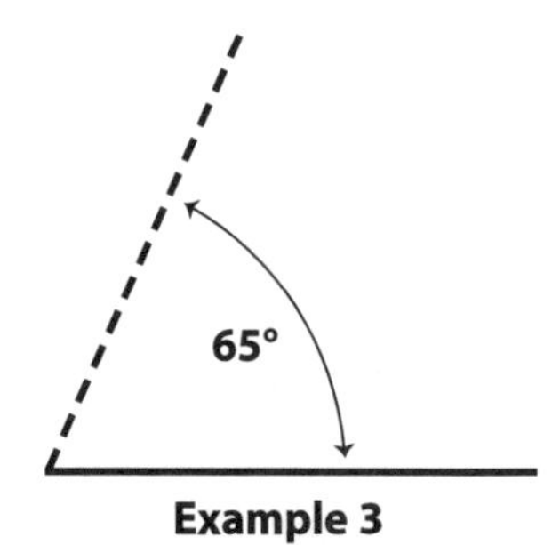

Example 3

Name ______________________________ Date ______________

Angular Measurements (continued)

Data and Results

Part I

A. 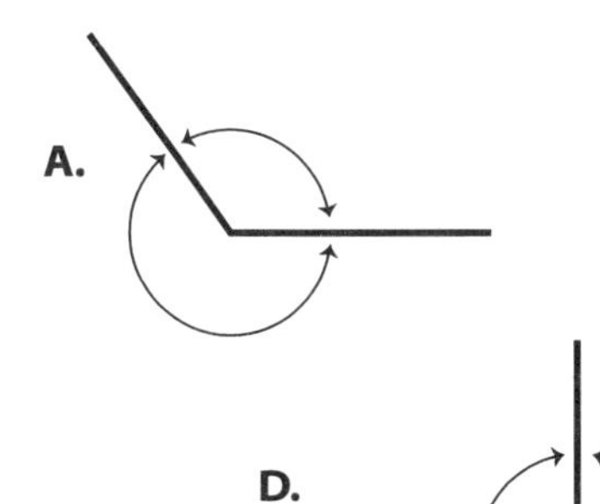

B. 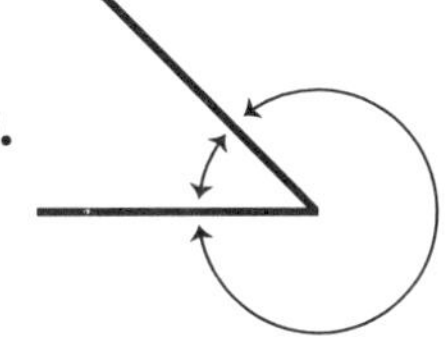

C. 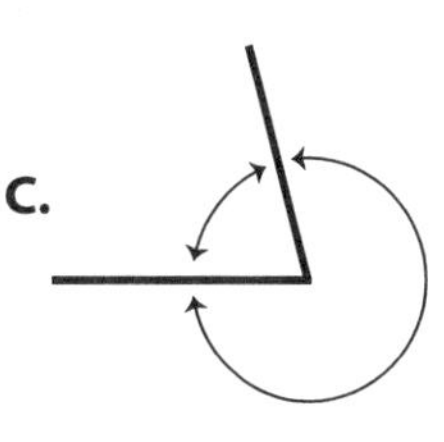

D.

E.

F. 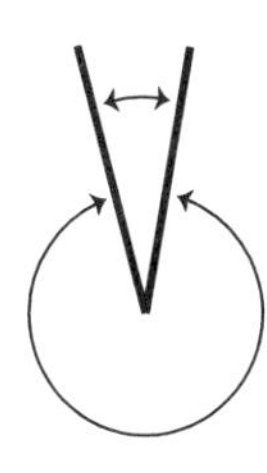

Part II

45°

G.

190°

H.

130°

I.

185°

J.

85°

K.

280°

L.

Part III

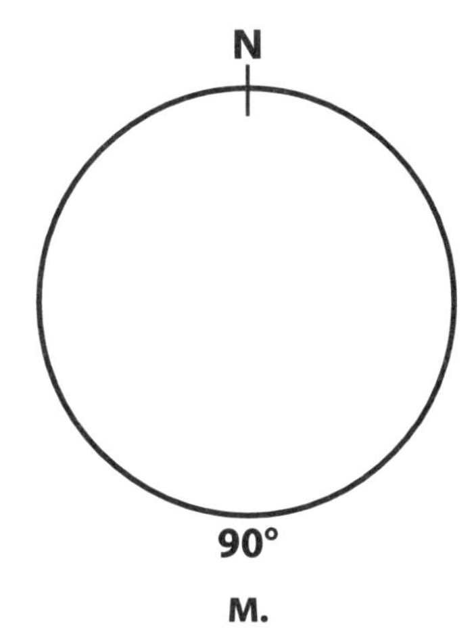

90°

M.

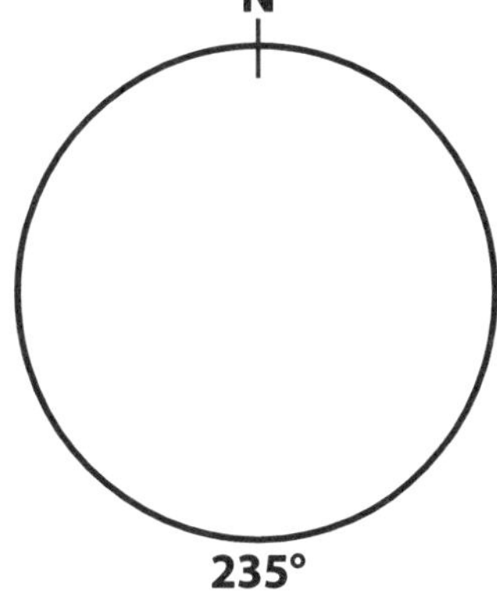

235°

N.

N

20°

O.

Name ______________________________ Date ______________

Mapping and Triangulation

Purpose: Measure and calculate the angles of a triangle.

Background: Scientists use triangulation to calculate distance when direct measurement is impossible or impractical, such as when measuring the distance across uneven terrain or across a body of water. Triangulation is based on a simple geometric rule that pertains to triangles: If you know the length of one side of a triangle and two adjacent angles, there is only one triangle you can construct from this information. By measuring a baseline and the angles from both ends of the baseline to a particular point, you can construct a triangle that can be used as a scale model of the area you are trying to measure.

Materials

- protractor
- calculator

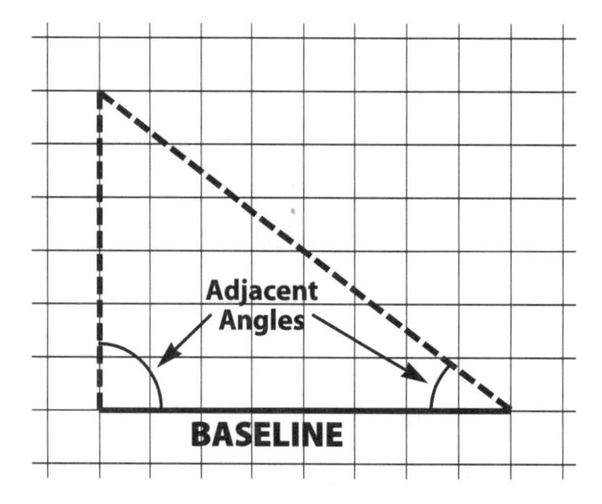

Procedure: Read the entire procedure before you begin the activity.

1. Look at triangles A–D in the "Data and Results" section. Use your protractor to measure the three interior angles of each triangle. Write the value inside the corresponding angle. Then add together the values of the three angles of each triangle, and write the total degrees next to the letter representing that triangle.
2. Look at triangles E–H in the "Data and Results" section. Using your calculator instead of a protractor, determine the degrees of the missing angle of each triangle. (Hint: Look at your results from step 1.) Write each answer inside the corresponding angle.
3. Look at line I in the "Data and Results" section. Use your protractor to draw at both ends of the baseline the angles that correspond to the measurements given. Your final result should be a triangle.

Name ______________________________ Date ______________

Mapping and Triangulation (continued)

Data and Results

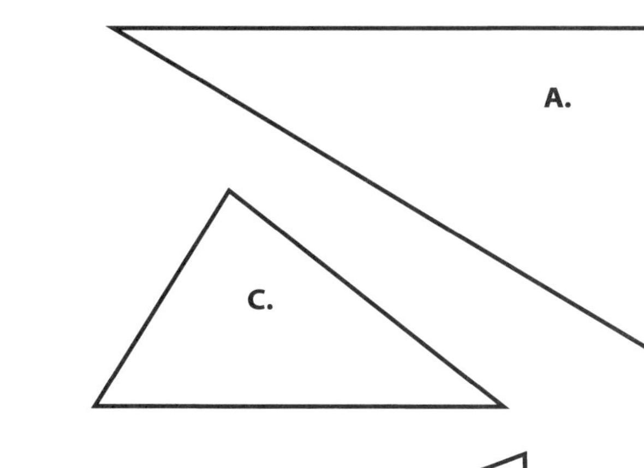

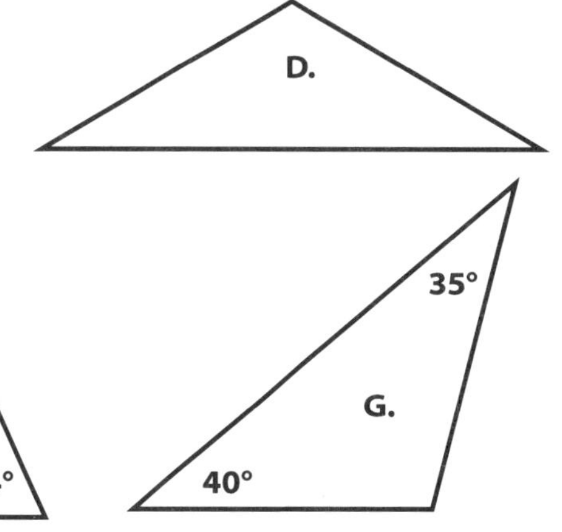

70°
E.
20°

F.
67°
64°

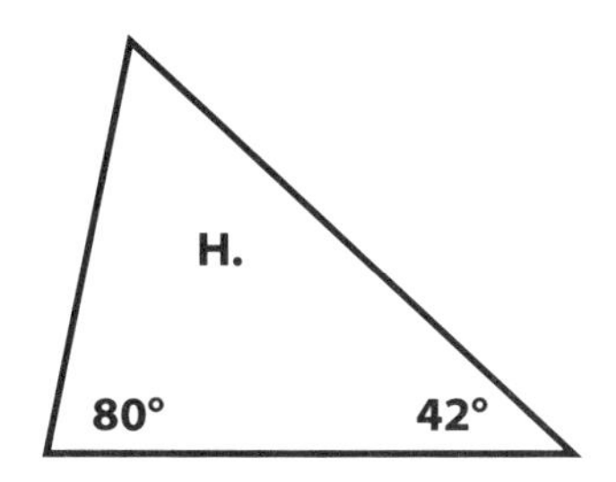

I.

Follow-Up

1. How did you determine the angular measurements for triangles E–H?

____

2. How many degrees do all three angles total in a triangle? _________

3. Write a detailed definition of a triangle.

____

____

Name ______________________________ Date ______________

Triangulation on a Scale Map

Purpose: Use triangulation to plot a position on a map.

Background: The technique of triangulation allows a person to calculate distances in difficult terrain. The continent of India, for example, was first mapped by using triangulation. This technique is based on knowing the distance of one side of a triangle, the baseline, and the two adjacent angles. If you know these three values, the entire triangle can be constructed to scale, and the length of the two remaining sides can be calculated.

Materials

- protractor
- metric ruler

Procedure: Before you begin the investigation, look at the map in the "Data and Results" section. Imagine that you have drawn this scale map while visiting Lake Terrapin during your vacation. The residents of the area have told you that a small island, which they call "Turtle Island," has been spotted out in the lake. To mark Turtle Island in the proper location on your map, read the following information and complete each step.

1. Your tent is at point A on the map, and you want this spot to represent one end of your baseline. You walk a straight path along the shoreline to point B—a large, flat rock. The distance between points A and B is 140 m. Use this information to draw a baseline on the map. Label the distance of the baseline (in meters).
2. Next, you measure the angle from point A to Turtle Island and find it to be 45°. Then you determine that the angle from point B, the flat rock, to Turtle Island is 30°. Use a protractor and your knowledge of triangulation to mark the location of Turtle Island on the map. Mark the spot with an *X*, and write the name of the island next to the *X*.
3. Finally, you want to know the actual distance (in meters) from your tent to Turtle Island and from the rock to Turtle Island. The scale of your map is 1 cm:10 m. Use a metric ruler and the information on your map to help you determine the actual distance to Turtle Island from both locations. Record this information on the map.

Name ______________________________ Date ______________

Triangulation on a Scale Map (continued)

Data and Results

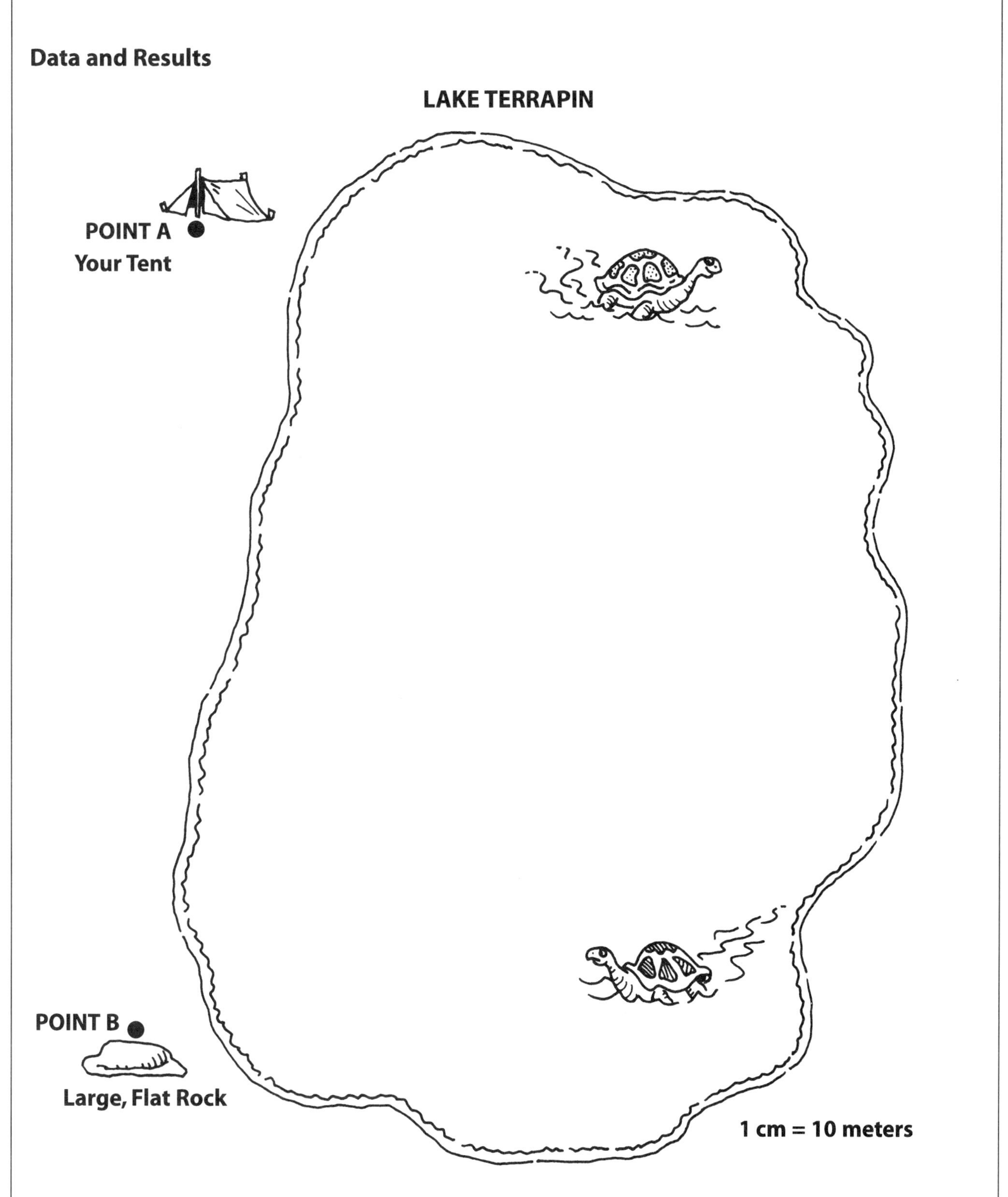

Name ______________________________ Date ______________

Triangulation on a Scale Map (continued)

Follow-Up

1. Explain how you knew where to plot the location of Turtle Island.

2. How far is it from Turtle Island to the tent? to the rock?

3. Suppose you had your map with you when you traveled on a boat to Turtle Island. A mist formed along the shore, and you were no longer able to see your camp, but you could see the large flat rock at Point B. How could you use a compass or a protractor to find your camp again?

4. How could you use triangulation to determine the height of a flagpole? Explain the process.

Name ______________________________ Date ____________

Using Right Triangles to Measure Height

Purpose: Determine the height of a flagpole by using a right triangle as the primary measuring tool.

Background: You have learned that triangulation is one method that can be used to measure distances, but another simple process involving proportions can also be used to determine the height of objects. This strategy requires the use of a simple right triangle as the primary measuring tool.

Materials

- 5" x 7" (12.5 cm x 17.5 cm) index card
- scissors
- small spirit level
- calculator
- tape
- meterstick

Procedure: Work with a partner. Read the entire procedure before you begin the investigation.

1. Cut a right triangle from the index card so that the sides that form the 90° are both 5" (12.5 cm) long. (Note that the index card is 5" x 7" [12.5 cm x 17.5 cm].)
2. Tape the paper triangle in an upright position to the spirited level so that a 45° angle is at one end and a 5" (12.5-cm) side is resting flat atop the level. (See Figure A).

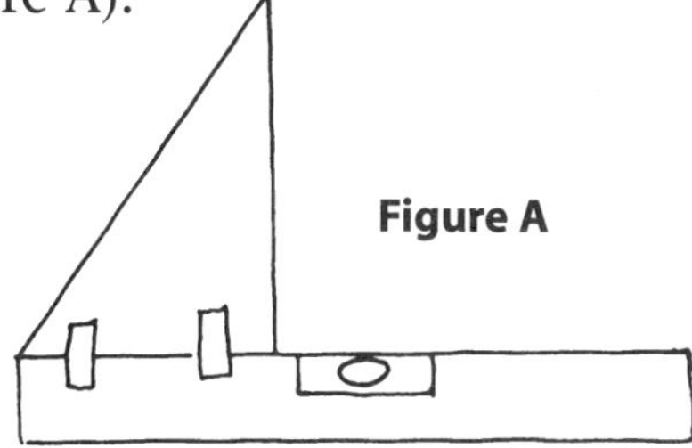

Figure A

3. Go outside to a flagpole and hold the triangle to your eye, with the 90° angle pointing toward the flagpole. Look up the long side of the triangle to see the top of the flagpole. Step forward or backward (be careful) until the top of the flagpole is in line with your eye and the long edge of the triangle. (See Figure B).

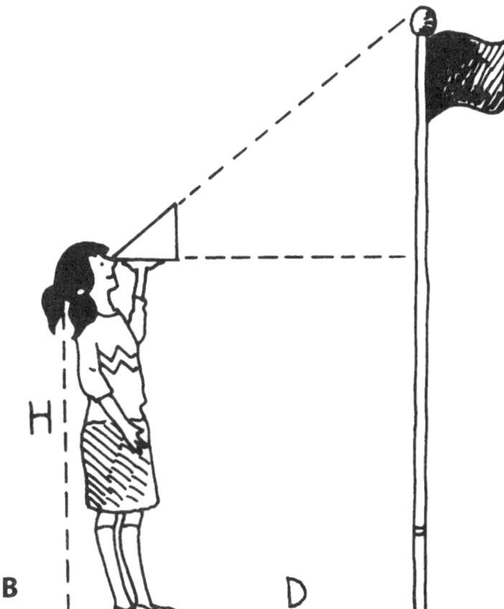

Figure B

Name ______________________ Date __________

Using Right Triangles to Measure Height (continued)

4. To make sure you are holding the triangle in the correct position, parallel to the ground, have your partner look at the bubble inside the spirit level to see if it is centered. If it is not, adjust your position until it is.
5. As you stand with the spirited level in the proper position, your partner should measure and record the height from the ground to your eyes (length h) and from where you stand to the flagpole (length d).
6. Add lengths h and d; the sum should equal the height of the flagpole. Record the values in the "Data and Results" section, under the header *Measurement 1.*
7. Switch roles with your partner, and repeat the procedure to get a second set of measurements. Record the values in the "Data and Results" section, under the header *Measurement 2.*
8. Measure the height of two other items (such as a telephone pole and the side of a building) by following the same procedure. Record your measurements and compare results.

Data and Results

Item	Measurement 1			Measurement 2		
	h	d	h + d	h	d	h + d
1. flagpole						
2. ______						
3. ______						

1. What two values did you get for the height of the flagpole? Were your measurements the same? Why or why not?

   ____

2. How do your values compare to those of your classmates? Are the results the same? Why or why not?

   ____

3. Explain why the height of the flagpole is equal to h + d.

   ____

   ____

Name ______________________________ Date ______________

Triangulation in Real Life

Purpose: Use the process of triangulation to measure distances in the classroom.

Background: The triangulation method can help you determine distances in real-life situations when direct methods of measurement are unavailable. See the practical advantages of triangulation by using the technique to determine the distance to various items in your classroom.

Materials

- adding-machine tape or butcher paper taped to worktable
- meterstick
- protractor
- straw
- graph paper

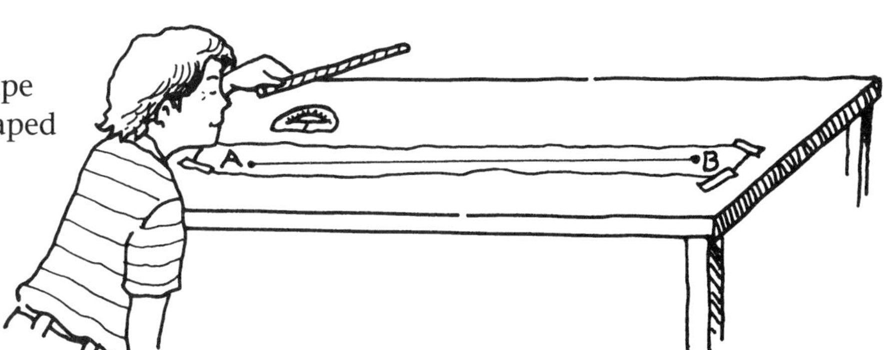

Procedure: Read the entire procedure before you begin the activity. Note that you should complete the activity *Triangulation on a Scale Map* before doing this one.

1. Use your meterstick to draw a 100-cm baseline on the paper taped to your worktable. Make a small reference dot at each end of the baseline, and label one dot *A* and the other dot *B*.
2. You will be using triangulation to measure the distance from your drawn line to different objects in the classroom. Ask your teacher to which objects you should measure, and then list these objects in the Measurements Table of the "Data and Results" section.
3. Place the protractor on dot A so that the crosshair or center point is exactly on the dot, and the protractor is oriented perpendicular to the baseline. Place the straw so that one end is directly above dot A, and then look through the straw to view one of the listed objects. Look at where the straw is aligned on the protractor, and measure the angle. Record the measurement in the data table under the header *Angle from A*.
4. Repeat step 3, but this time view the item from dot B. Record the measurement in the data table under the header *Angle from B*.
5. Repeat steps 3 and 4 for each item listed in the data table.

Name ______________________________ Date ______________

Triangulation in Real Life (continued)

6. Use the information you've collected to help you draw scale triangles that represent the distance to each item. (Hint: The baseline and the two angles you've measured for each item make a triangular shape that can be drawn to scale, such as 1 graph square: 10 cm).
7. Use each scale triangle to determine the distance to the item from points A and B. (Hint: You will need to use your scale ratio to convert values from the scale length to the actual distance.) Record your values in the data table, under the headers *Distance from A* and *Distance from B.*

Data and Results

MEASUREMENTS TABLE

Item	Angle from A	Angle from B	Distance from A	Distance from B
1. ______	______	______	______	______
2. ______	______	______	______	______
3. ______	______	______	______	______
4. ______	______	______	______	______
5. ______	______	______	______	______
6. ______	______	______	______	______
7. ______	______	______	______	______
8. ______	______	______	______	______

Name ______________________________ Date ____________

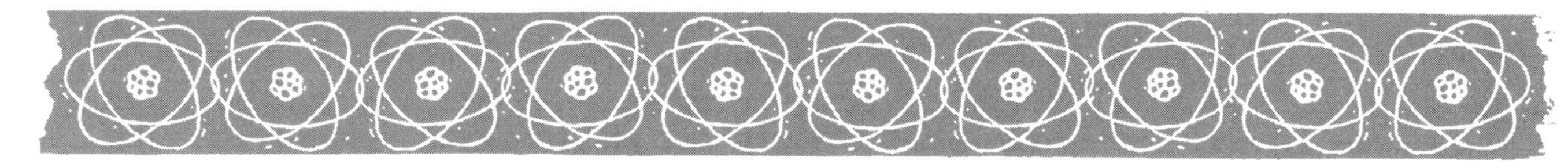

Mathematical Model of Radioactive Decay

Purpose: Simulate the radioactive decay of three different "elements."

Background: Certain radioactive elements, such as uranium, give off particles and energy as they decay into other lighter elements. This radioactive decay takes place at a constant, measurable rate that can be used to determine the age of rocks and minerals containing radioactive elements. Different radioactive isotopes decay at different rates. The half-life is the amount of time needed for half of the atoms of a given mass of substance to decay.

Materials

- square- or rectangular-shaped box with lid
- cup of uncooked rice
- marker
- masking tape
- graph paper

Procedure: Work with a partner. Read the entire procedure before you begin the activity. Note that in the following decay simulation, each grain of rice represents an atom.

1. Count out 100 grains of rice. Use a marker to color one end of each grain of rice.
2. Use masking tape to label the sides of the box *A*, *B*, *C*, and *D*. Then place the rice inside the box and cover it with the lid.
3. Hold the lid in place as you pick up the box and tap it once on the bottom to shake up the "atoms." Then set the box down and slowly take off the lid so as not to disturb the positioning of the rice grains. Carefully remove each grain that has its colored end pointing toward side A (consider these grains as having just "decayed"). Then count the number of grains you removed from the box.
4. Calculate the number of grains that remain inside the box by subtracting the number of grains removed from the starting amount (in this case, 100). Record this value (the number of grains left in the box) as *Element A, Shake 1* in the "Data and Results" section.
5. Repeat steps 3 and 4, and record the amount of remaining rice as *Element A, Shake 2*. (Note that the starting amount for the calculation is no longer 100, but the amount that was in the box after Shake 1). Continue this process (for Shakes 3, 4, 5, and so on) until all the "atoms" have "decayed" (there is no more rice in the box).

Name ______________________________ Date ______________

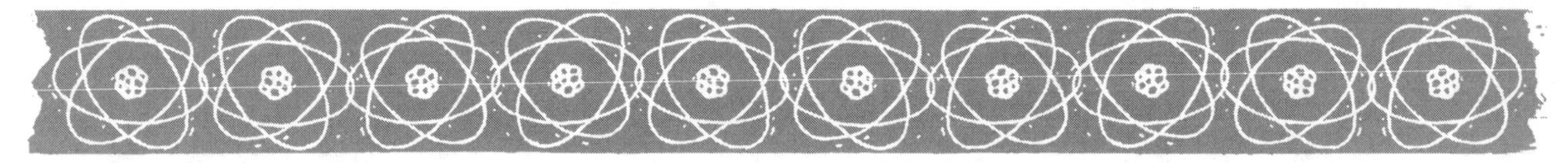

Mathematical Model of Radioactive Decay (continued)

6. Place the 100 grains back inside the box. (Note: Recount the grains to make sure you have them all.) Repeat the entire process, but this time remove the grains that point toward either side A or side B (those pointing toward sides C and D stay inside the box). Record the results under *Element B* in the data table.
7. Repeat step 6, but this time remove the grains that point toward side A, side B, or side C. Record the results under *Element C* in the data table.
8. Graph your results (Shake Number vs. Amount of Rice Grains That Remain) to show the decay rate of each element. Be sure to label each line.

Data and Results

DECAY TABLE
(Remaining Rice Grains)

Shake	Element A	Element B	Element C
1			
2			
3			
4			
5			
6			
7			
8			
9			
10			
11			
12			
13			
14			
15			
16			

Name ______________________________ Date ______________

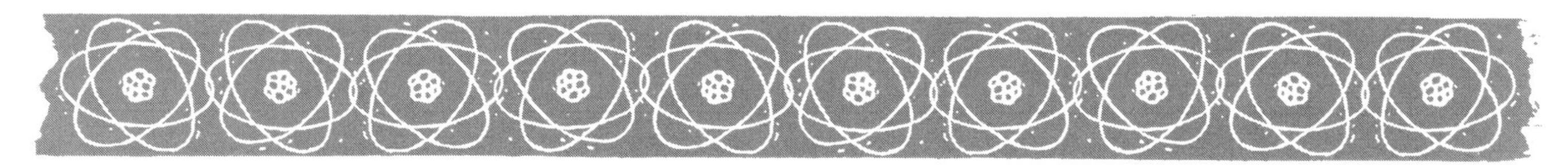

Mathematical Model of Radioactive Decay (continued)

1. Determine the half-life of each element (the number of shakes required for half of the "atoms" to decay).

   Element A: __________ Element B: __________ Element C: __________

2. Compare your results with those of your classmates. Are all the half-lives the same?

   ____

3. Using the class results, what is the average half-life of Element A? Element B? Element C?

   Element A: __________ Element B: __________ Element C: __________

4. Was the box you used square or rectangular? How would this affect your results?

   ____
   ____
   ____
   ____
   ____
   ____
   ____

5. How random do you think the decay rate was in this simulation? What were some of the variables that may have affected the randomness of the experiment?

   ____
   ____
   ____
   ____
   ____
   ____
   ____

Name ______________________________ Date ____________

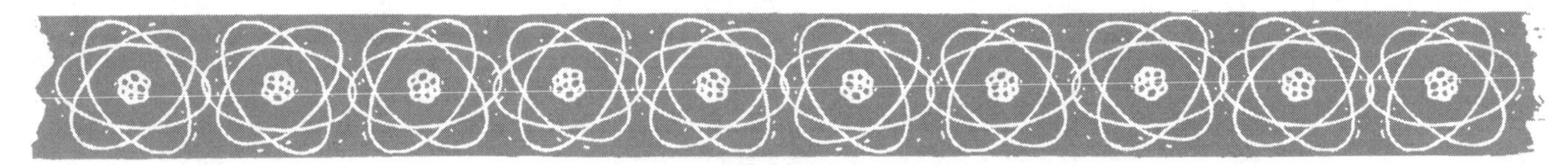

Melting Ice

Purpose: Measure and analyze the melting rate of ice.

Background: The Celsius scale (°C) is universally used for scientific measurement of temperature, whereas the Fahrenheit scale (°F) is usually used for engineering and household purposes. When analyzing a collection of temperature readings, scientists may calculate the mean (average) and the mode (value that occurs most frequently).

Materials

- foam cup
- water at room temperature
- Celsius thermometer
- ice cubes
- clock with a second hand
- sponges or paper towels
- graph paper

Procedure: Work with two classmates. Read the entire procedure before you begin. As you complete this investigation, never leave the thermometer unattended, as it may tip out or be knocked over and break.

1. Assign group roles. Decide who will be the **recorder**, the **thermometer monitor,** and the **group leader.**
2. **Group leader:** Fill a foam cup ⅔ full with water, and then place a thermometer in the cup.

   Thermometer monitor: Use the thermometer to gently stir the water for 30 seconds. Then read the temperature without lifting the thermometer from the water.

   Recorder: Write the noted temperature in the data table, next to number 1.
3. **Group leader:** Add two ice cubes.

   Thermometer monitor: Use the thermometer to gently stir the ice water for 30 seconds. Then read the temperature without lifting the thermometer from the water.

   Recorder: Write the noted temperature in the data table, next to number 2.

Name ______________________________ Date ______________

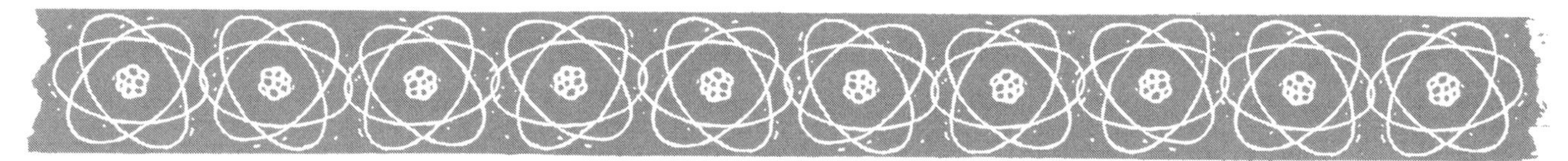

Melting Ice (continued)

4. Repeat step 3, recording temperature readings for numbers 3, 4, 5, and so on, until your group gets the same temperature reading four times in a row. Keep track of how many ice cubes are added and how much time elapses. (Note that it may be necessary to pour out small amounts of water from time to time to prevent overflow. The group leader is responsible for pouring out the water and cleaning up any spills.)
5. After the temperature remains constant, do not add any more ice. Continue to read and record the water's temperature every 30 seconds until 48 readings have been taken. Rotate jobs every 16 readings so that each group member has the opportunity to practice reading a thermometer. Remember: Do not take the thermometer out of the water until all the data is recorded.
6. After all the data has been collected and recorded, graph your results.

Data and Results

Temp. (°C)	Temp. (°C)	Temp. (°C)	Temp. (°C)
1. ________	13. ________	25. ________	37. ________
2. ________	14. ________	26. ________	38. ________
3. ________	15. ________	27. ________	39. ________
4. ________	16. ________	28. ________	40. ________
5. ________	17. ________	29. ________	41. ________
6. ________	18. ________	30. ________	42. ________
7. ________	19. ________	31. ________	43. ________
8. ________	20. ________	32. ________	44. ________
9. ________	21. ________	33. ________	45. ________
10. ________	22. ________	34. ________	46. ________
11. ________	23. ________	35. ________	47. ________
12. ________	24. ________	36. ________	48. ________

Name ____ Date ______________

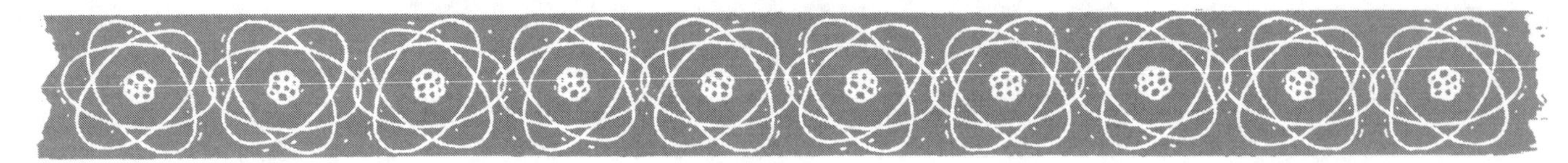

Melting Ice (continued)

1. What was the lowest temperature measured? _______
2. What was the highest temperature measured? _______
3. What is the difference between the highest and lowest temperature? _______
4. How long did it take for the ice water to reach its lowest temperature? _______
5. How many ice cubes were added before the ice water reached its lowest temperature? _______
6. What is the difference in temperature between temperature #1 and the lowest temperature? _______
7. What is the difference in temperature between temperature #48 and the lowest temperature? _______
8. Before the temperature reached its lowest point, how long did it take for the temperature to drop
   5 degrees? _______ 10 degrees? _______
9. After the temperature reached its lowest point, how long did it take for the temperature to rise
   2 degrees? _______ 4 degrees? _______ 6 degrees? _______
10. What is the mode of the temperatures you recorded? _______
11. What is the mean of the temperatures you recorded? _______
12. Compare your readings to those of other groups. What was the lowest temperature measured? _______
13. Compare your readings to those of other groups. What was the highest temperature measured? _______
14. What is the difference between the highest and lowest temperature measured in the class? _______
15. How does your group's results compare to the answers written for questions 12–14? Explain why there may be a difference in results.

    ____

    ____

Name ______________________________ Date ______________

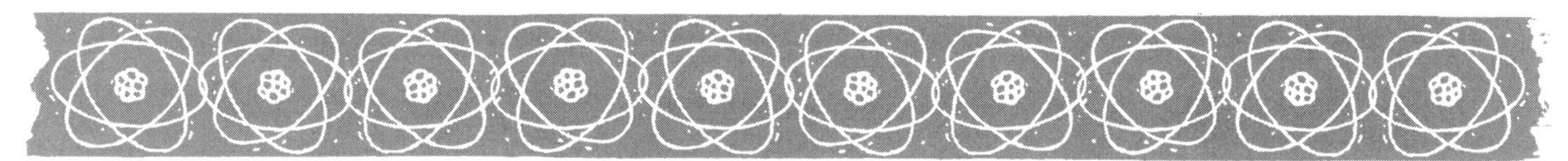

Vapor Escaper

Purpose: Determine the relationship between the evaporation rate and the surface area of water.

Background: Heat energy from the sun causes water on Earth's surface to evaporate in a process known as the water cycle. The rate of evaporation of water can be studied in mathematics as an example of a quadratic function as well as an example of how numbers relate to nature's energy sources.

Materials

- 3 empty, metal food cans of different sizes (all cylindrical shape)
- masking tape
- marker
- metric ruler and/or calipers
- water
- triple-beam balance or electronic scale
- graph paper

Procedure: Read the entire procedure before you begin the investigation.

1. Make sure all three cans are clean and dry. Then use tape to label the cans of increasing diameter *A*, *B*, and *C*.
2. Determine the radius (½ of the diameter) of each can, and record the value in the data table of the "Data and Results" section.
3. Measure the mass of each can, and record the value in the data table.
4. Add water to can A to within 1 cm of the top. Weigh the water-filled can, and calculate the mass of the water by subtracting the mass of the empty can from the total (water's mass = mass of water-filled can – can's mass). Record the value in the data table. Then set the can aside without spilling any water.
5. Repeat step 4 with cans B and C. Record the results in the data chart.
6. Place each can in the same location, exposed to the exact same conditions. Try to make sure that none of the cans are under direct sunlight or that one is closer to a heat source or window than the others. Every day for four days, measure the mass of water in each can, and record the values in the data table. Note: The measurements should always be taken at the same time of day (e.g., 10:00 A.M.).
7. Graph your results to show the evaporation rate for each can (Mass of Water vs. Time).

Name ______________________________ Date ______________

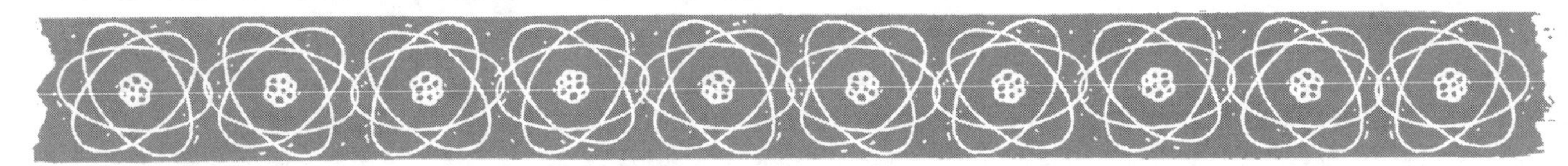

Vapor Escaper (continued)

Data and Results

DATA TABLE

Mass of Water

Can	Radius	Mass of Can	Day 1 (start)	Day 2	Day 3	Day 4
A						
B						
C						

1. Calculate the percentage of water loss for each can per day (water's mass/starting mass x 100). Which can lost the highest percentage of water during the experiment?

   DAY 2: Can A _______ Can B _______ Can C _______

   DAY 3: Can A _______ Can B _______ Can C _______

   DAY 4: Can A _______ Can B _______ Can C _______

   Which can lost the most water? __________

2. Calculate and record the surface area of each can ($A = 3.14 \times r^2$). Which can had the greatest surface area of water exposed to the air?

   Surface Area of A: _______ Surface Area of B: _______ Surface Area of C: _______

   Which can had the most water exposed to air? __________

3. What is the relationship between evaporation rate and surface area?

   ____

   ____

   ____

   ____

Name ______________________________ Date ______________

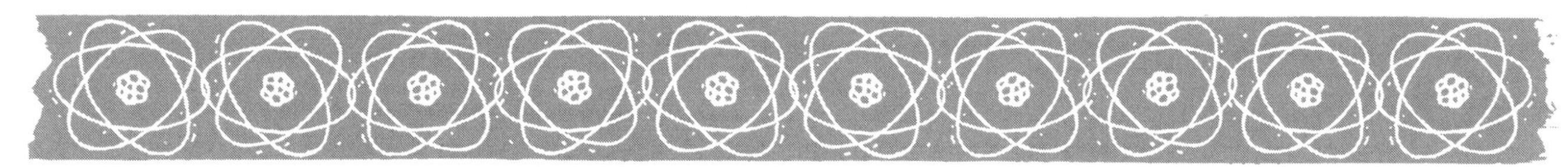

Turn Up the Heat

Purpose: Determine how color affects the heating and cooling rates of surfaces.

Background: The heating and cooling rates of different surfaces depend on several factors. Composition of the surface material, angle to the sun, texture, exterior temperature, wind flow, and humidity are factors that influence these rates. But there is another, perhaps more important factor: color.

Materials

- 2 sheets of different-colored construction paper
- foam board or cardboard
- 4 rubber bands
- 2 Celsius thermometers
- 100-watt lightbulb in a clamp-light
- ring stand
- textbooks
- laboratory tools (scissors, metric ruler, protractor, tape, calculator)
- writing paper
- graph paper
- colored pencils

Procedure: Work with three classmates. Read the entire procedure before you begin the investigation. Note that your teacher will assign you the two colors to test in this activity.

1. Cut a 8½" x 11½" (21 cm x 28 cm) rectangle from each colored paper and from two pieces of foam board or cardboard.
2. Stack the following items in the order listed: cardboard or foam-board rectangle, colored rectangle, thermometer (laid flat and lengthwise across the rectangular pieces). Make two stacks—one for each colored rectangle. Use two rubber bands to bind each stack of items. (Wrap the rubber bands tightly so the thermometer doesn't slide out.)

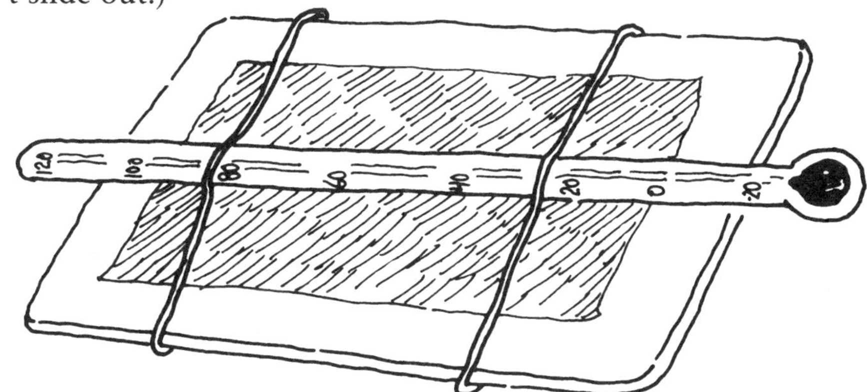

Name ____________________ Date ____________

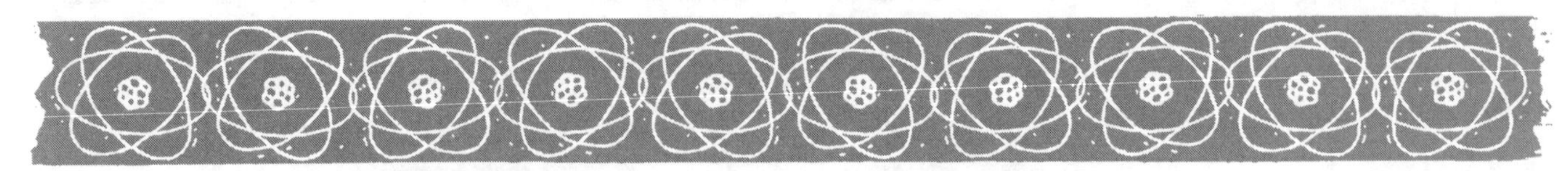

Turn Up the Heat (continued)

3. Attach the clamp-light to the ring stand so that the center of the lightbulb is 20 cm above the tabletop. Do not turn on the light yet!
4. Position both thermometer packs on the tabletop so they are 50 cm from the center of the lightbulb and at a 60° angle. Use a protractor to help you angle the thermometer packs, and use textbooks and tape to prop up and secure the packs in place. Double-check to make sure that the colored rectangles, the thermometers, and the clamp-light are all correctly positioned. Then record in the Heat Rate Table the colors your group is testing.

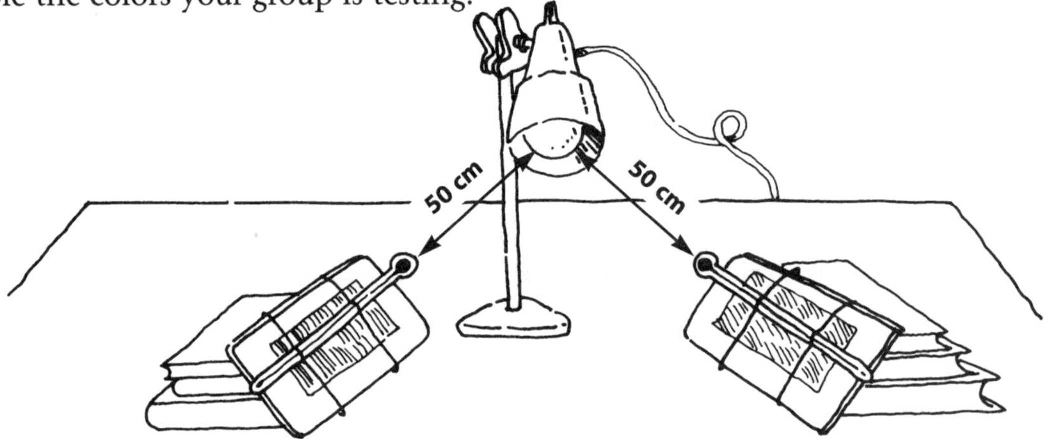

5. As a group, decide who will read the thermometers, who will monitor the time, and who will record the data. Note that group members should switch roles half-way through the testing so that everyone gets practice reading a thermometer.
6. Read the starting temperatures (°C) on the thermometers, and record each value in the appropriate column of the table, next to *Time 0.* (Note that you will be calculating °F after you have collected the data.)
7. Turn on the lightbulb and begin testing. Read and record the temperatures of the thermometers every minute. After 10 minutes (10 readings), turn off the light and continue reading and recording the temperatures every minute for another 10 minutes.
8. After 20 minutes (20 readings), stop testing. Use the formula *°F = (°C x 1.8) + 32* to convert each of your recorded measurements from Celsius to Fahrenheit. Record the converted values in the appropriate spaces of the Heat Rate Table.
9. Compare your results to those of other groups. On one sheet of paper, graph the class results for each color test (Time vs. Temperature,°F). Use corresponding colored pencils to draw each line graph (for example, a red-colored pencil to draw the results of red paper).

Name ____________________ Date __________

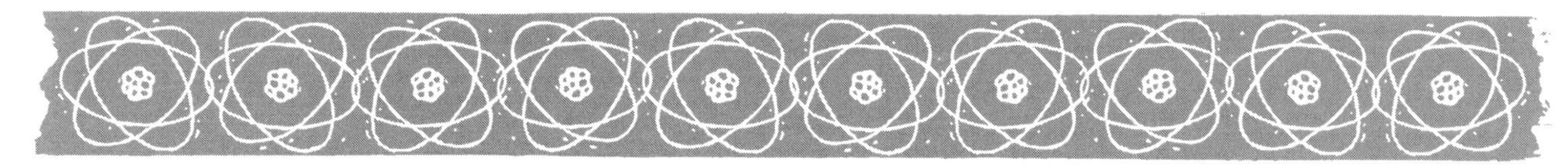

Turn Up the Heat (continued)

Data and Results

HEAT RATE TABLE

	________ COLOR TESTED		________ COLOR TESTED	
Time (minutes)	**Temp. (°C)**	**Temp. (°F)**	**Temp. (°C)**	**Temp. (°F)**
0				
1				
2				
3				
4				
5				
6				
7				
8				
9				
10				
(turn off the light) 11				
12				
13				
14				
15				
16				
17				
18				
19				
20				

Name ______________________________ Date ______________

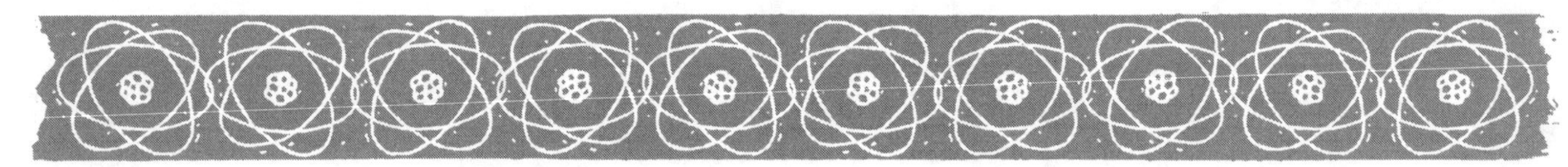

Turn Up the Heat (continued)

1. Which color absorbed the most heat? ______________

2. Which color absorbed the least amount of heat? ______________

3. Which color released the most heat (cooled off) after the light was turned off?

   ____

4. Which color released the least heat after the light was turned off? ______________

5. Why was the temperature taken before the light was turned on?

   ____

   ____

   ____

6. What variables, other than color, may have influenced your results?

   ____

   ____

   ____

   ____

7. Did more than one group test the same color? If so, were the results the same? Why or why not?

   ____

   ____

   ____

   ____

   ____

   ____

Name ______________________________ Date ______________

Stretching the Truth

Purpose: Explore the relationship between the amount of weight attached to a rubber band and the distance the rubber band stretches.

Background: A *function* in mathematics represents a relationship between two variables that are dependent on each other. For example, the speed of an automobile is dependent (in part) on the amount of gasoline being used.

Materials

- 2 metersticks
- large paper clip
- string
- scissors
- 12 washers
- graph paper
- rubber band

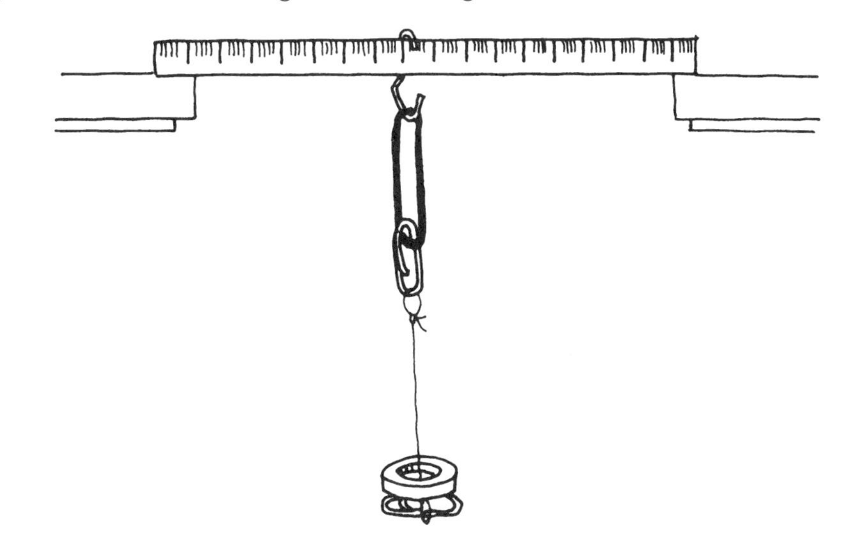

Procedure: Read the entire procedure before you begin the activity. Refer to the above diagram as you complete each step.

1. Place a meterstick between two tables or chairs to make a support beam. Bend a paper clip into a hook, and hang it from the meterstick. Then hang a rubber band on the hook.
2. Make a loop at both ends of a piece of string so that the string's length is 4" (10 cm). Attach a paper clip to the bottom loop, and then slip a washer onto the string so that it rests on top of the paper clip (and doesn't slide off the string).
3. Attach another paper clip to the top loop, and then hook it to the bottom of the hanging rubber band.
4. Measure the length of the elongated rubber band, and record the measurement in the data table.
5. Remove the string, add another washer, and repeat the measuring process. Continue adding washers and taking measurements until you've filled in the data table.
6. Graph your results (Number of Washers vs. Rubber-Band Length).

Name ________________________ Date ____________

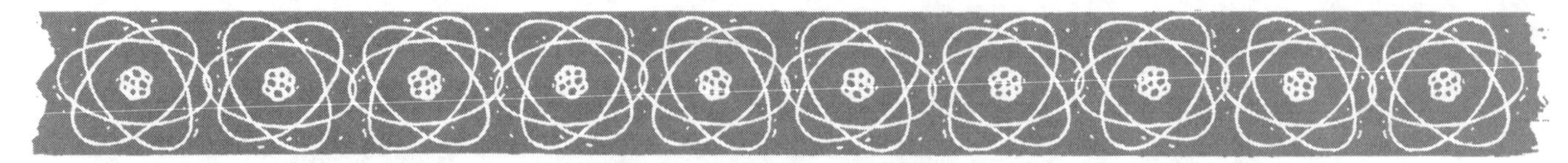

Stretching the Truth (continued)

Data and Results

# of Washers	Rubber-Band Length (cm)	# of Washers	Rubber-Band Length (cm)
1	________	6	________
2	________	7	________
3	________	8	________
4	________	9	________
5	________	10	________

Follow-Up

1. What are the variables in this activity? What is the difference between a dependent variable and an independent variable?

 ____

 ____

2. What remained constant?

 ____

3. Express the relationship between the number of washers and the length of the rubber band.

 ____

 ____

4. This activity demonstrates a linear relationship. What does that mean?

 ____

5. What are some other examples of functions?

 ____

 ____

Extension

Repeat the activity with another rubber band that is a different size. Compare your results.

Name ______________________________ Date ______________

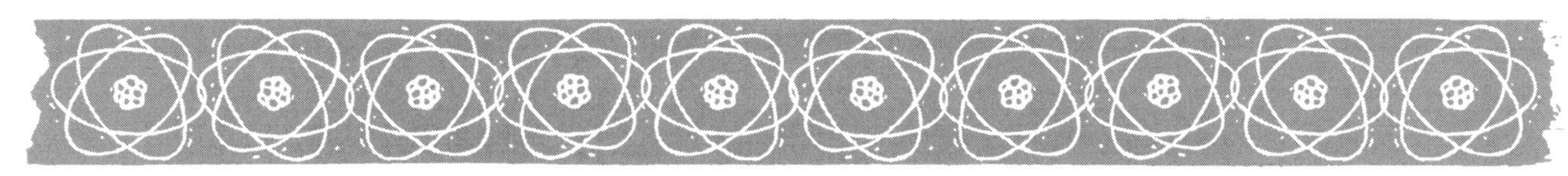

Follow the Bouncing Ball

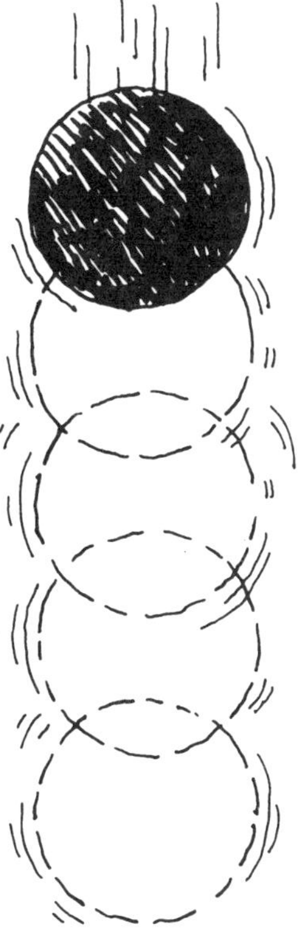

Purpose: Construct and use a bounce chamber to measure the "decay rate" of a bouncing ball.

Background: Radioactive isotopes decay at a predictable rate, as demonstrated in the activity *Mathematical Model of Radioactive Decay*. A bouncing ball also follows decay-like behavior when it is dropped. Each successive bounce is lower and lower until the ball no longer leaves the surface.

Materials

- 3 different types of balls
- string
- meterstick
- stiff poster board
- scissors
- pencil
- markers
- clear plastic wrap
- clear tape
- video equipment
- graph paper

Procedure: Work with two classmates. Read the entire procedure before you begin the investigation. Note that you will be using video equipment to tape and then view how high a ball bounces.

1. Use string and a meterstick to measure the circumference (length C, in cm) of the largest ball you will be testing. Then cut out a poster-board strip that has a length of 1 m and a width of (3C + 3) cm.
2. With a pencil, draw two parallel lines down the length of the strip to divide it into three sections, each of which is (C + 1) cm wide. (See Figure A.)

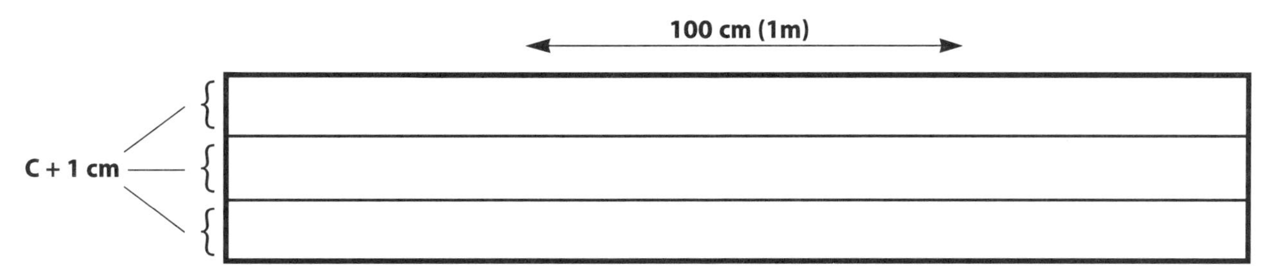

FIGURE A

Name ______________________ Date ____________

Follow the Bouncing Ball (continued)

3. Use a marker and a meterstick to draw lines width-wise across the paper to mark off a scale from 1–100 cm. (See Figure B.) Number off each increment of the scale (for example, every 20 cm), and darken the lines in the center so they are clearly visible when looked at through a video camera placed three meters away.

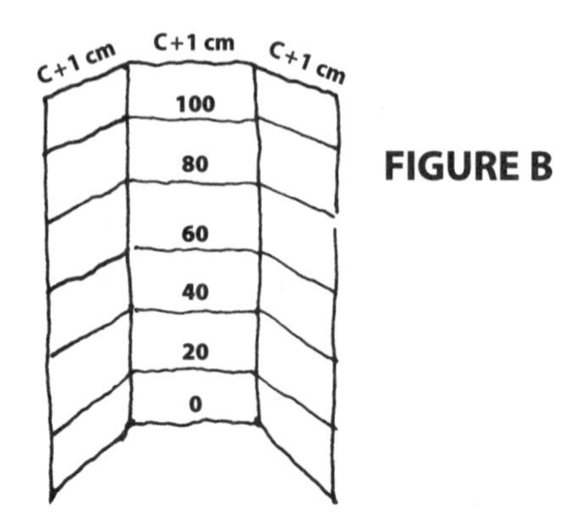

FIGURE B

4. Construct a three-sided trough by folding the poster board lengthwise, alongside each pencil-drawn line. (See Figure C. Note that the fourth side of the trough is open-faced). Test the trough to make sure the largest ball can roll through it.

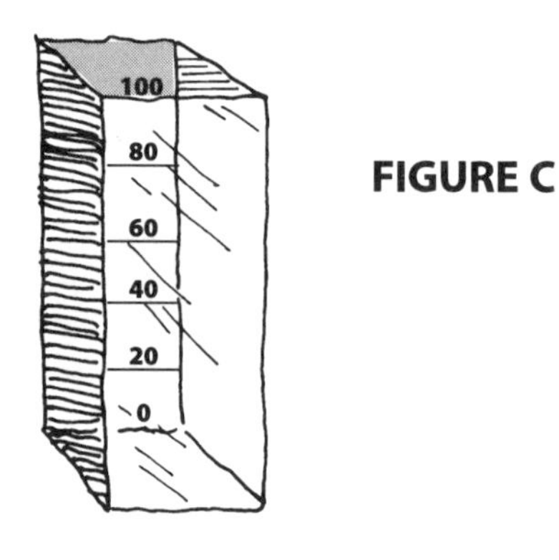

FIGURE C

5. Cover the open side of the trough with clear plastic wrap. Then stand up the trough so that the plastic-covered side is facing forward and the scale reads *0–100* from bottom to top. (See Figure C.) This "bounce chamber" will keep the ball from moving too much in a horizontal direction.

6. Set the video camera in front of the bounce chamber, at a distance in which both the scale and the top of the chamber can be clearly seen in the viewfinder. (See Figure D.) Try to situate the camera so that it avoids glare from the plastic wrap. Test the setup to make sure that your camera positioning is correct. Note that the camera position should be quite low (about 100 cm from the floor).

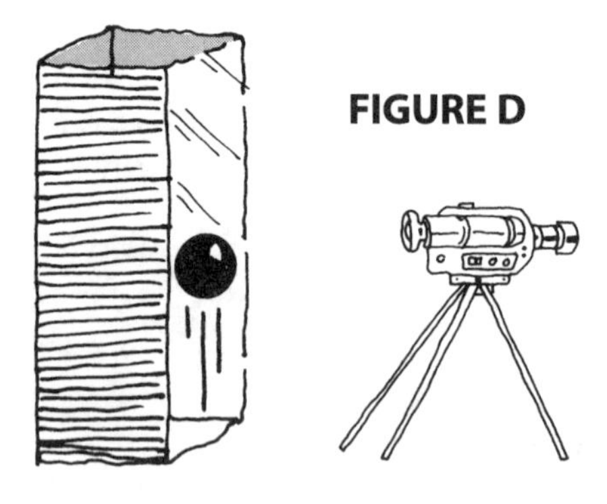

FIGURE D

7. Once the setup is confirmed, begin testing. Have one person in your group hold onto the bounce chamber to keep it steady, another person monitor the camera (being careful not to move it), and the last person drop the ball through the top of the chamber at the 100-cm mark. Videotape the ball in motion until it stops bouncing.

8. Remove the ball from the chamber, and repeat the experiment two more times with two other balls. (Note: Make sure not to erase the taping of any previous tests.)

Name ______________________________ Date ______________

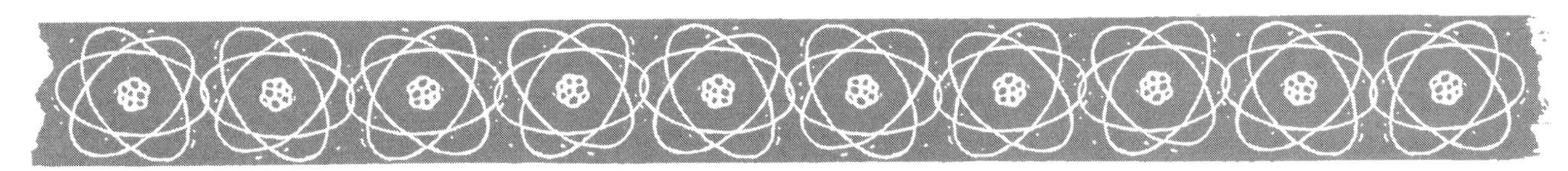

Follow the Bouncing Ball (continued)

9. After you have completed all testing, view the videotape in a VCR. Use the freeze-frame function to confirm the height of each bounce. Rewind and replay each test as needed to ensure accuracy. In the "Data and Results" section, record the number and the corresponding measurement of each bounce of the ball. (Note that each ball will most likely bounce a different number of times.)
10. On one sheet of paper, graph your results for each ball to show its "Decay Rate" (Bounce Number vs. Height).

Data and Results

1st Ball	2nd Ball	3rd Ball
Bounce # ____ **Height** _______	**Bounce #** ____ **Height** _______	**Bounce #** ____ **Height** _______
Bounce # ____ **Height** _______	**Bounce #** ____ **Height** _______	**Bounce #** ____ **Height** _______
Bounce # ____ **Height** _______	**Bounce #** ____ **Height** _______	**Bounce #** ____ **Height** _______
Bounce # ____ **Height** _______	**Bounce #** ____ **Height** _______	**Bounce #** ____ **Height** _______
Bounce # ____ **Height** _______	**Bounce #** ____ **Height** _______	**Bounce #** ____ **Height** _______
Bounce # ____ **Height** _______	**Bounce #** ____ **Height** _______	**Bounce #** ____ **Height** _______

Follow-Up

1. Which ball had the shortest decay rate? the longest? Explain your answers.

____

____

2. Why was a bounce chamber constructed?

____

____

3. What variables may have affected this experiment?

____

____

4. Why does a ball bounce?

____

____

Name ______________________________ Date ____________

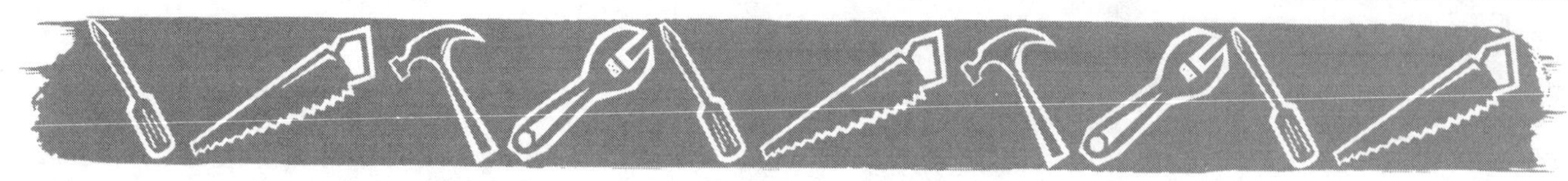

The Tallest Tower

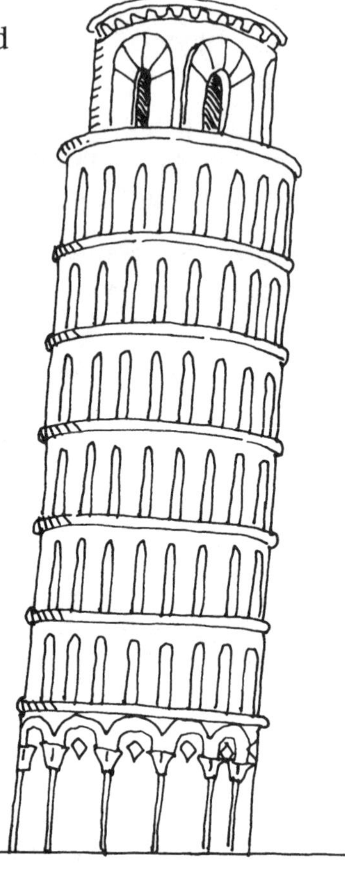

Purpose: Construct a tall tower that is able to withstand the forces of a strong wind.

Background: Building a tall, structurally strong but lightweight tower is a challenge for engineers. Many factors have to be considered, such as design, materials, soil conditions, climate, and expense.

Materials

- 25 straws
- 100 straight pins
- electric fan (1 per class)

Procedure: Work with a partner. Read the entire procedure before you begin construction. Note that you will be making a tower that will be tested in a competition for height and durability.

1. Your goal is to use straws and straight pins to build the tallest tower possible that will withstand the wind from a fan. Plan how you will accomplish this feat. Then write your plan in the "Data and Results" section.
2. Build your tower. Note that the tower must be freestanding—not pinned or attached to anything, such as carpeted floor. The straws may be bent or folded; however, the pins must be kept straight. No other alterations are allowed.
3. When you've finished building your tower, move it to the "test area" that your teacher has designated. Then observe as your teacher turns on the fan to test the sturdiness of your building. If your tower tips over or loses a piece, it will be disqualified.
4. With each round of the competition, your teacher will increase the intensity of the wind by moving the fan closer or by increasing its speed. The last tower to remain standing and intact will be declared the winner. If there is a tie, the tallest tower will win.

Name ______________________________ Date ______________

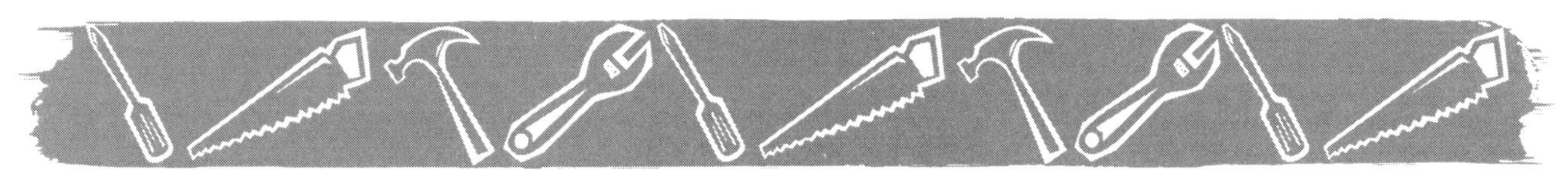

The Tallest Tower (continued)

Our Plan:

____

____

____

____

____

____

1. Draw a sketch of the winning design.
2. How well did your building withstand the wind? If you could construct the tower again, how would you improve your design?

   ____

   ____

3. What features did the best designs have that allowed them to withstand higher wind velocities?

   ____

   ____

4. What characteristics did some of the less successful designs seem to have?

   ____

   ____

5. If you could replace the pins with another material to hold the straws together, what would that material be and why?

   ____

   ____

Name ______________________________ Date ______________

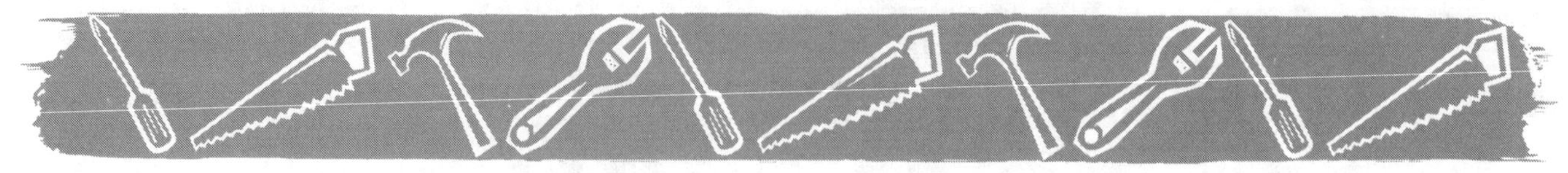

The Strongest Bridge

Purpose: Construct a sturdy bridge capable of supporting increasing amounts of weight.

Background: A bridge is a lifeline across a river or a canyon. In today's cities, it is also a pathway upon which traffic can pass through a congested area. When designing a strong bridge, engineers must take into consideration factors that may cause the bridge to collapse, such as decades of heavy traffic, environmental pollution, and extremes in climatic conditions.

Materials

- 25 drinking straws
- 100 straight pins
- metric ruler
- string
- scissors
- standard weights (1 set per class)
- stopwatch (1 per class)
- pillow (1 per class)

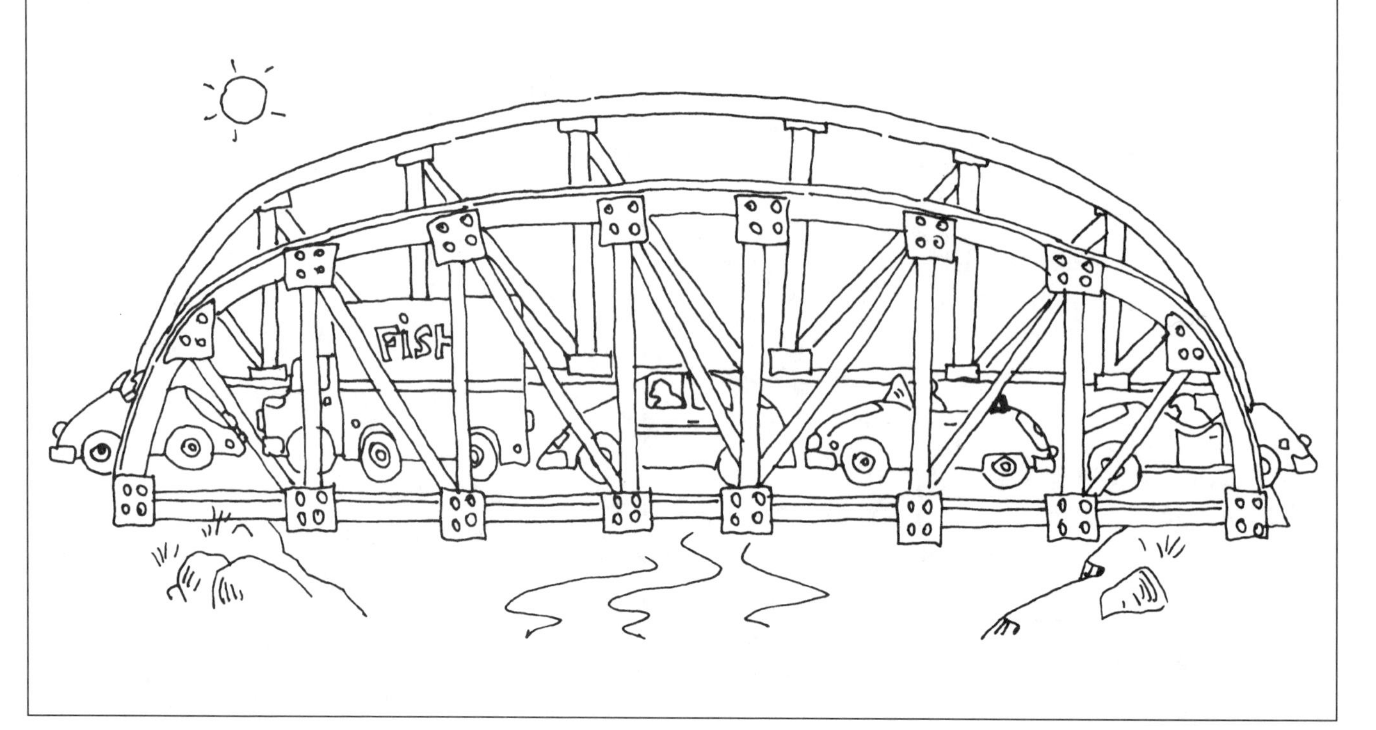

Name ______________________________ Date ______________

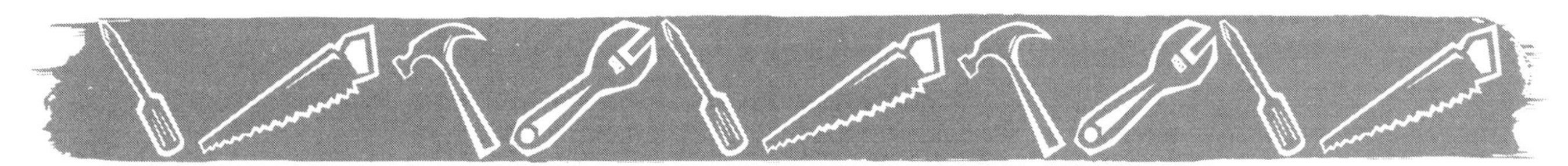

The Strongest Bridge (continued)

Procedure: Work with a partner. Read the entire procedure before you begin construction. Note that you will be making a bridge that will be tested in a competition for strength and sturdiness.

1. Your goal is to use straws and straight pins to build a sturdy bridge long enough to span a 20-cm gap between two tables. Plan how you will accomplish this feat. Then write your plan in the "Data and Results" section.
2. Build your bridge. Note that the straws may be bent or folded; however, the pins must be kept straight. No other alterations are allowed.
3. As directed by your teacher, place the bridge between two tables that are spaced 20 cm apart. Once the bridge is in place, you may not reposition it (unless it is accidentally moved from its original position).
4. Tie a loop at the bottom of a piece of string so that the total length of the string is 30 cm. (You may cut the string as needed.) When it is your turn to test your bridge, complete the following steps:

   a. Attach the string to the center of the bridge so that the loop hangs freely at the bottom.

   b. Hook one weight onto the loop. Note that the suspended weight should not be touching the floor or any object below the bridge.

   c. After each 30-second interval, add another weight to the loop. Your teacher and classmates will help you keep track of the total amount of weight being added.

   d. When your bridge finally collapses, determine its strength—the weight it supported for at least 30 seconds before it fell apart.

5. The bridge that supports the most weight before collapsing will be declared the winner.

Name ______________________ Date ____________

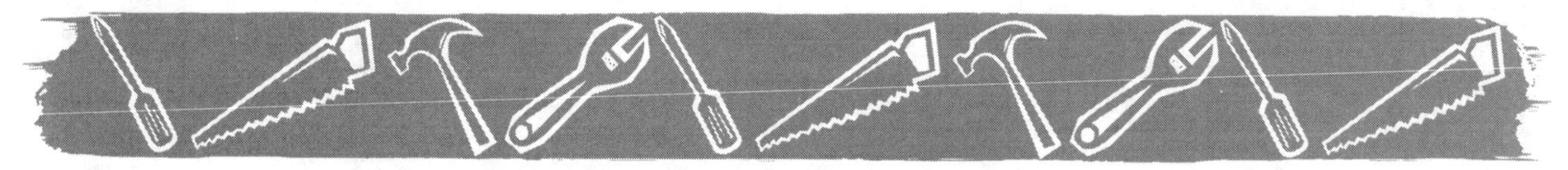

The Strongest Bridge (continued)

Data and Results

Our Plan:

1. What was your bridge's strength (the total amount of weight it supported)?

2. What features did the best bridge designs in class have that allowed them to support the most weight?

3. How would you design your bridge differently? Draw a sketch of your revised bridge.

Name ______________________ Date __________

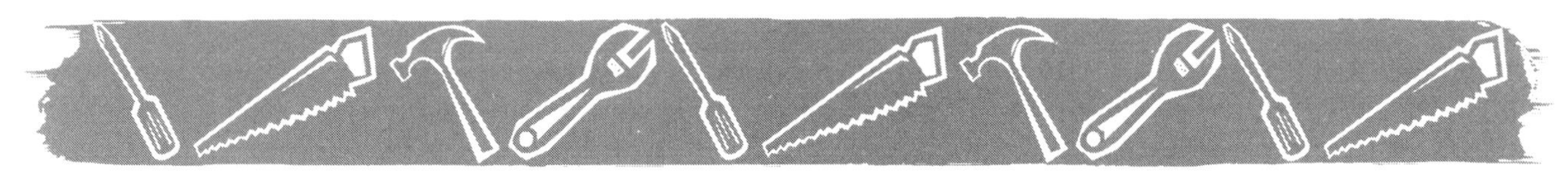

Oops! The Egg Dropped

Purpose: Construct protective packaging for a raw egg that is being dropped from increasing heights.

Background: Some engineers design special packaging that protects items from being damaged or destroyed when moved or stored. The development of these special "cushioning" materials have allowed the transport of extremely fragile objects such as glass, electronic parts, and eggs.

Materials

- 25 drinking straws
- 100 straight pins
- raw chicken egg

Procedure: Work with a partner. Read the entire procedure before you begin construction. Note that you will be making an "egg protector" that will be tested in a competition.

1. Your goal is to use straws and straight pins to build a container that will protect an egg that is dropped from increasing heights. Plan how you will accomplish this feat. Then write your plan in the "Data and Results" section.
2. Build the container around an egg. Note that the straws may be bent or folded; however, the pins must be kept straight. No other alterations are allowed.
3. After the class has finished building their containers, your teacher will take you and your classmates to a designated drop site. On command, test your structure by dropping it onto the plastic sheet. If the egg breaks, cracks, or comes out of the container, it will be disqualified. If it survives the fall, it should be given to your teacher for further testing. (Note: Clean up any mess you make.)
4. Observe your teacher as he or she tests the surviving eggs (hopefully one of them being yours) by dropping them from progressively higher levels. The encased egg that survives the highest drop will be declared the winner. (Note that your teacher may declare multiple winners if more than one egg survives the highest drop.)

Name ______________________________ Date ______________

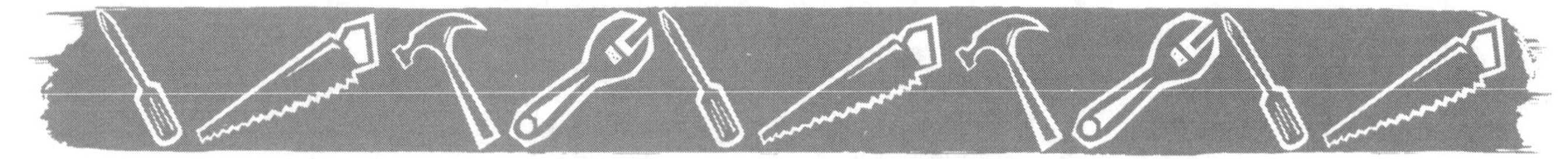

Oops! The Egg Dropped (continued)

Data and Results

Our Plan:

1. Make a sketch of the winning design.

2. What features did the most successful containers have that allowed them to protect the enclosed egg at a greater height?

3. How well did your egg withstand the falls? If you could construct the container again, how would you improve your design?

4. How would you test real-life packaging materials such as foam peanuts, bubble wrap, and newspaper? Explain the steps you would perform.

Name ______________________________ Date ______________

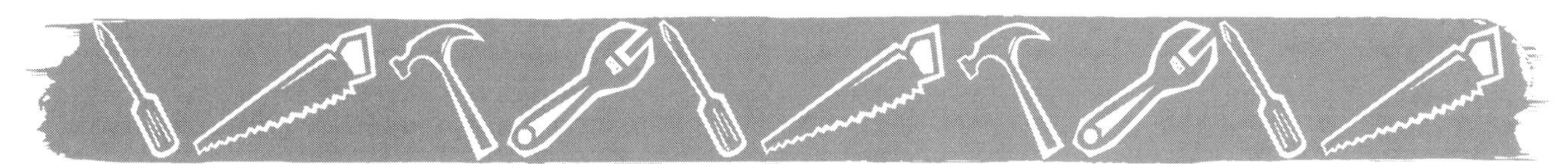

First-Rate Flying

Purpose: Construct a paper airplane that can travel long distances and remain in the air for long periods of time.

Background: Many airplanes are capable of flying extremely long distances and staying in the air for many hours. Some have even circled the globe nonstop. Their distance and duration is limited by the amount of fuel they are able to carry and by pilot fatigue.

Materials

- sheet of white copy paper
- scissors
- tape measure or meterstick (1 per class)
- stopwatch (1 per class)

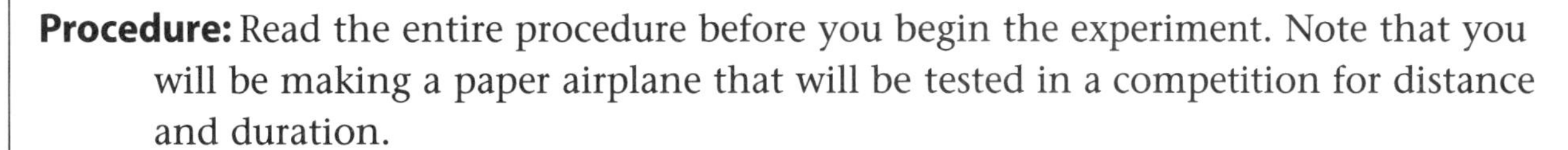

Procedure: Read the entire procedure before you begin the experiment. Note that you will be making a paper airplane that will be tested in a competition for distance and duration.

1. Using only one sheet of paper, construct a paper airplane that will glide the greatest distance from a launch point. You may use a pair of scissors if you need to cut or shape the paper, but you may not use tape, glue, or any other binding material. Note that your aircraft must have recognizable wings and be capable of gliding. A wadded up or tightly folded piece of paper will be unacceptable and disqualified from the competition.
2. After the class has finished making their airplanes, your teacher will take you and your classmates to a launch site. When instructed to do so, launch your airplane. Observe the airplane in flight as your teacher times how long it stays in the air. When your airplane hits the ground, use a tape measure or meterstick to measure the distance it traveled. Record your results in the "Data and Results" section.
3. If time permits, test your airplane again, and take the average of your results to determine how far and how long your airplane traveled. Your teacher will announce two winners—one for distance and the other for duration.

Name ______________________________ Date ______________

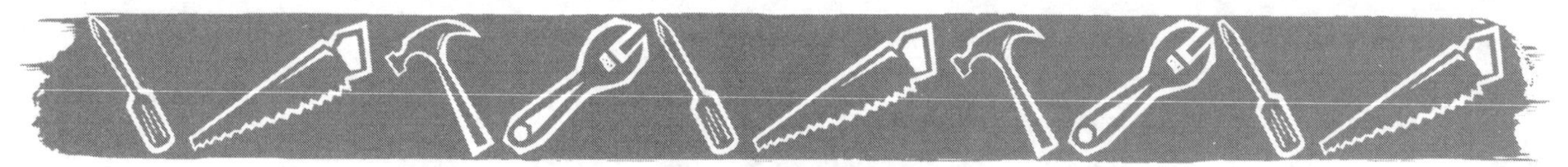

First-Rate Flying (continued)

Data and Results

Distance Traveled: ______________ Time in the Air (Duration): ______________

What design elements enabled the winning airplanes to travel farther or longer than the other airplanes?

____

____

____

____

____

____

____

____

____

____

Glossary of Terms

Absolute magnitude: A measure of a star's brightness. It is based on a comparison with the sun's illumination, which is assigned a brightness value of 1.

Angstrom: A unit of length equal to one ten-billionth of a meter.

Barometer: A device for measuring barometric pressure (air pressure).

Celestial body: Referring to planets, stars, the moon, the sun, and other heavenly bodies.

Celsius scale (°C): A scale for measuring temperature on which 0 degrees marks the freezing point of water and 100 degrees marks the boiling point.
$°C = (°F - 32)/1.8$

Density: The amount of mass per unit volume ($d = m/v$). It is a measure of the compactness or thickness of an object or substance.

Fahrenheit scale (°F): A scale for measuring temperature on which 32 degrees marks the freezing point of water and 212 degrees marks the boiling point.
$°F - 32 = (1.8)(°C)$

Function: The relationship between two variables that are dependent on each other.

Half-life: The time required for the disintegration of one-half of the radioactive atoms in a sample.

Humidity: The amount of moisture or water vapor in the air.

Isotopes: Two or more atoms of the same element that have different atomic masses due to different numbers of neutrons.

Luminosity: The quality or condition of being bright or full of light.

Mass: The quantity of matter in a particular object or substance. The mass of an object is always the same—on Earth, on the moon, in outer space—whereas its weight, which depends on the force of gravity, can vary.

Mean: The average value of a set of numbers. For example, the mean of numbers 3, 3, 4, and 6 is the sum of the numbers divided by 4 (since there are four numbers being averaged): $(3 + 3 + 4 + 6)/4 = 4$.

Metric system: A decimal system of weights and measures used by scientists and mathematicians worldwide.

Mode: The number that occurs most frequently in a set of numbers. For example, the mode of numbers 3, 3, 4, and 5 is the number 3 because that value occurs most often.

Natural selection: A process in nature by which living things that are well adapted to their environment tend to survive and reproduce more often than those that are not.

Parallax: The apparent change in the position of an object when it is seen from two different points not on a straight line with the object. It is used in surveying and astronomy to determine the distance of objects.

Period: The time it takes for a pendulum to complete one full swing.

Radioactive decay: The process by which an atomic nucleus disintegrates, spontaneously emitting energetic particles (as electrons or alpha particles). The result is a transmutation of an element, or a change from one element to another.

Relative humidity: A ratio that compares the amount of water vapor present in the air to the greatest amount it could contain at the same temperature. Relative humidity is often expressed as a percentage.

Scale: A representation (as in a map, model, or figure) of an actual object, drawn according to a set ratio. For example, a scale of 1 cm:1 km means that every centimeter of the model represents one kilometer of the actual distance.

SI: Abbreviation for the International System of Units. A standard system of metric measurements used by mathematicians and scientists worldwide.

Significant figures: The number of significant digits expressed in a measurement, which indicates how precise the measurement is. For example, the number 34,700,000 has three significant figures, which means that the true value of the measurement is between 34,650,000 and 34,750,000. The last significant figure (e.g., the 7) is the doubtful digit. Comparatively, the measurement 34.700 has five significant figures.

Specific gravity: The ratio of the density of a substance to the density of an equal volume of pure water (usually at 4°C). In metric units, the density of water at 4°C is 1.000 gram/cm^3.

Triangulation: A method of using triangles and scale drawings to determine distance.

Variable: A quantity that may assume any one of a set of values, depending on experimental conditions.

Volume: The amount of space an object or substance occupies.

Weight: A measure of the amount of gravitational force that is pulling a body toward the center of Earth. Since weight is dependent on gravity, it can vary when measured in different parts of the universe (for example, on the moon). It is proportional to mass—the amount of matter in an object. Because this proportionality is essentially constant over the surface of Earth, weighing an object is equivalent to measuring its mass.

Answer Key

The actual values that students measure and record will vary in most of the activities. This section provides some of the expected results and the answers to questions posed at the end of the investigations.

A System of Standards (p. 25)

1. The nub measurements should be more accurate. It is a smaller unit of measure, which leaves less room for error.
2. Nub and stem values among students should vary because each person's finger length and forearm length are different.
3. The use of standard units enables researchers to more easily and accurately compare results.

What Unit Should I Use? (p. 26)

Across

2. gram
4. micrometer
5. milliliter
6. kilometer
7. liter
8. milligram

Down

1. centimeter
3. angstrom
5. millimeter
6. kilogram

Measuring Length (p. 28)

1. 0.32 dm, 3.2 cm, 32 mm
2. 0.26 dm, 2.6 cm, 26 mm
3. 0.53 dm, 5.3 cm, 53 mm
4. 0.95 dm, 9.5 cm, 95 mm
5. 0.42 dm, 4.2 cm, 42 mm
6. 0.74 dm, 7.4 cm, 74 mm
7. 0.86 dm, 8.6 cm, 86 mm
8. 1.15 dm, 11.5 cm, 115 mm
9. 0.14 dm, 1.4 cm, 14 mm
10. 0.33 dm, 3.3 cm, 33 mm
11. 0.66 dm, 6.6 cm, 66 mm
12. 1.22 dm, 11.2 cm, 112 mm
13. 1.03 dm, 10.3 cm, 103 mm
14. 0.09 dm, 0.9 cm, 9 mm
15. 1.51 dm, 15.1 cm, 151 mm

More Measurements

Answers will vary.

Measuring Volume and Density (p. 31)

1. Density is the mass per unit volume (d = m/v).
2. You need the object's mass and volume.
3. Use a balance or scale to measure the mass of a solid. To determine the mass of a liquid, weigh it in a container and then subtract the mass of the empty container.
4. The volume of the solid can be determined by measuring the dimensions and using a simple formula. The volume of liquid can be measured in a graduated cylinder.
5. Place the object in a graduated cylinder of water and measure how much the water level rises. The water displaced by an object is equal to its volume.

Measuring Specific Gravity (p. 33)

Answers will vary.

The Importance of Significant Figures (p. 37)

1. The metric ruler, because of its smaller increments.
2. The smallest graduated cylinder, because of the closer level of accuracy, possible.
3. The electronic scale; it's more sensitive if correctly calibrated.

Devising Measuring Strategies (p. 40)

Answers will vary.

Do You Overestimate or Underestimate Yourself? (p. 42)

Answers will vary.

Determining the Moon's Diameter (p. 44)

1. Answers will vary.
2. Answers will vary.
3. The moon's diameter is 3,476 km.
4. Possible errors include an inaccurate measuring device, bent or non-parallel cards, poor alignment of the moon's image, and errors in calculations.

Scale Models of Planets and Moons (pp. 46–48)

1. Paper tape is more practical for long-distance models.
2. Measure the distance between the two, and then use the scale to convert the value.
3. The scale models would be smaller in size, but still proportional.
4. Using a particular scale insures that the models are a practical size.
5. The model is the same size as the object it represents.
6. Practicality—ships are much larger than cars.
7. The model would not accurately represent the airplane; it would be a strange-looking representation with parts too small or too large by comparison.

Extension Activity: Answers for 1–5 will vary, depending on the accuracy of the model and subsequent calculations. The order of the planets, from smallest to largest, is Pluto, Mercury, Mars, Venus, Earth, Neptune, Uranus, Saturn, and Jupiter.

Star Light, Star Bright (p. 50)

1. The temperature increases as you move to the right.
2. The luminosity (brightness) decreases as you move from top to bottom.
3. A star in that region would be a hot and bright star.

Predicting Sunspots (pp. 53 and 54)

1. The average is about 11 years between sunspot maximums.
2. The average is about 11 years between sunspot minimums.
3. The greatest number of years between consecutive maximums is 12 years.
4. The fewest number of years between consecutive maximums is 10 years.
5. The greatest number of years between consecutive minimums is 12 years.
6. The fewest number of years between consecutive minimums is 8 years.
7. The next solar maximum probably occurred around 1990, 10 years after the maximum in 1980. The next solar minimum probably occurred around 1986, 10 years after the last solar minimum in 1976.
8. The last solar maximum probably occurred just before 1884, 10 years prior to the solar maximum in 1894.
9. The last solar minimum probably occurred around 1878, 10 years prior to the solar minimum in 1888.
10. Answers will vary.
11. Overall, the sunspot activity was relatively low during the time when radios were lowest in quality. Therefore, it had much less influence on the communications equipment than it would have if the activity was at very high levels.

How Much Humidity? (p. 57)

1. Answers will vary.
2. The relative humidity would decrease.
3. It is 100%, which means no evaporation takes place.
4. The measurement would not take into account the temperature of the air, which is also a variable.

Weather Watch (p. 60)

1. Low barometric pressure indicates cooler, windy, stormy weather; high barometric pressure indicates calm, warmer, stable weather.
2. Answers will vary.
3. The kind of weather moving in your direction might be a good answer.
4. Answers will vary.

Survival of the Fittest (p. 64)

1. Answers may vary, but usually the smallest rodents have the best survival rate.
2. Answer will vary.
3. Some factors include the ability to hide and the ability to run away from predators.
4. The rodents with the highest surviving populations will most likely produce the most.
5. Animals who can blend in with their environments are less likely to be spotted by predators.
6. Traits important in natural selection include reproductive rates, defensive and evasive abilities, and adaptations that are favorable to the environment (for example, long necks in an environment of tall fruit trees).

A Salty Solution (p. 66)

Answers will vary. Students should conclude that the measured values for the solution are less than the calculated values.

The Parallax Principle (pp. 68 and 69)

The values in the "Data and Results" section will vary, as will answers to the follow-up questions. Students most likely will design a triangular-shaped instrument with a sight (a device that aids the eye in aiming or in finding the direction of an object). Factors that can cause errors include misalignments and sight errors.

Angular Measurements (p. 71)

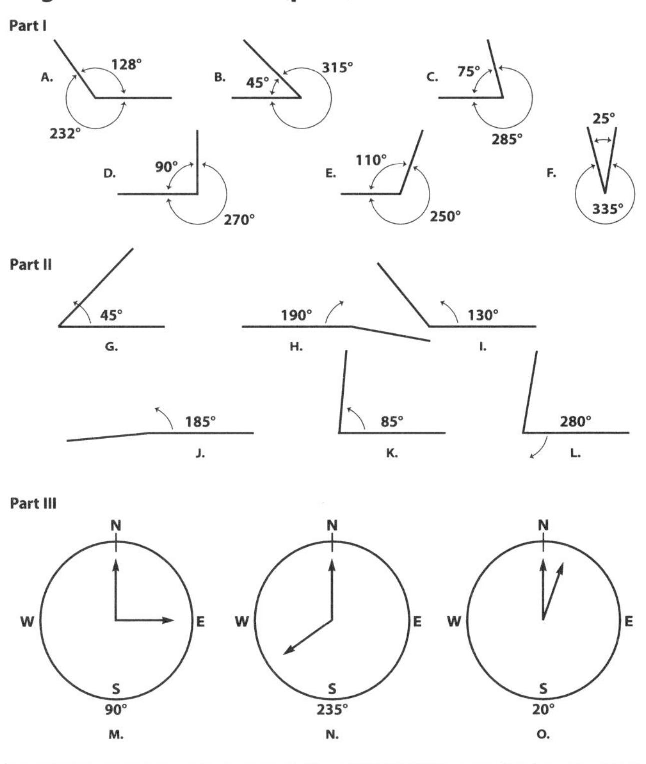

Mapping and Triangulation (p. 73)

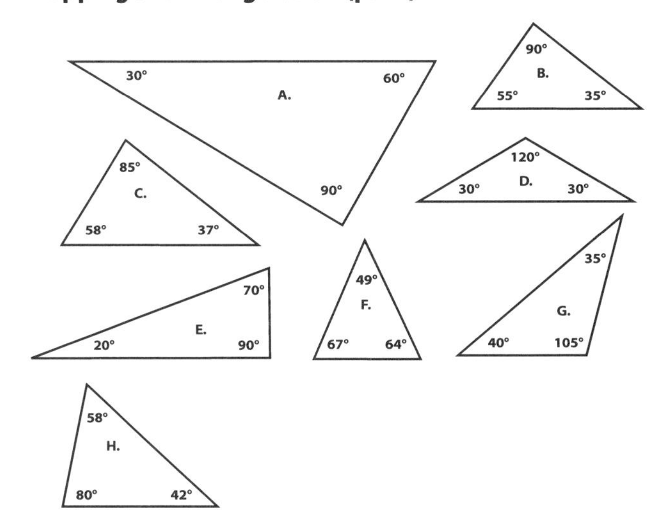

1. Add together the measured angles, and subtract the total from 180.
2. The sum of the angles of a triangle always equals 180°.
3. Possible answer: Three intersecting straight lines forming three interior angles that total 180 degrees.

Triangulation on a Scale Map (p. 76)

1. It is the point of intersection of the lines that extend from points A and B.
2. Answers will vary slightly, depending on measurement techniques. Students' answers should be close to:

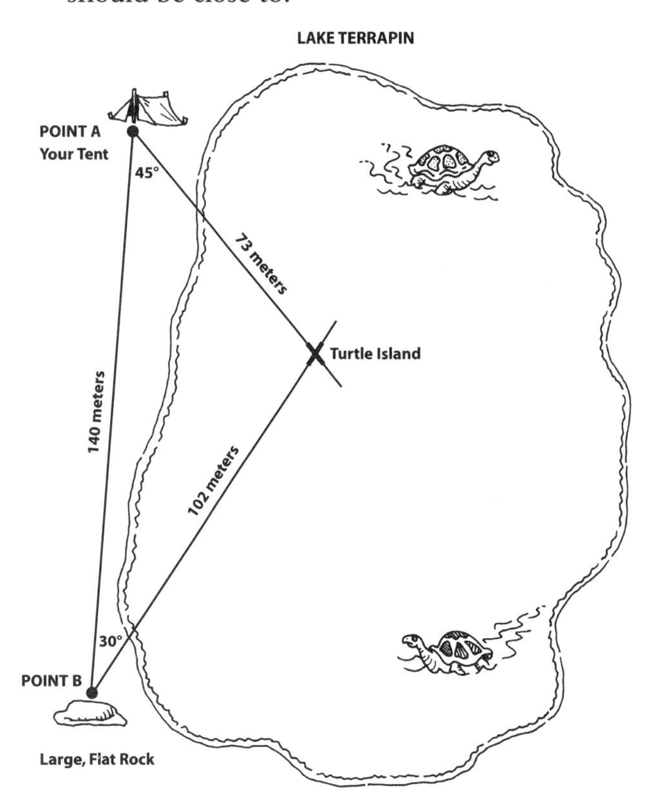

3. You could determine which angle (or heading) to travel.
4. Measure the angles vertically and duplicate the basic process.

Using Right Triangles to Measure Height (p. 78)

Measurements will most likely vary slightly, depending on how accurately students position the levels and measure the distances.

Triangulation in Real Life (p. 80)

Answers will vary.

Mathematical Model of Radioactive Decay (p. 83)

Answers will vary.

Melting Ice (p. 86)

Answers will vary.

Vapor Escaper (p. 88)

Answers will vary. Students should discover that the evaporation rate is directly proportional to the amount of water exposed to the air.

Turn Up the Heat (p. 92)

Answers will vary. Students should discover that dark colors absorb more heat than light colors.

Stretching the Truth (p. 94)

1. The variables are weight and stretching distance. A dependent variable (in this case, the amount of stretching) is one whose measurement is affected by an independent variable (in this case, the weight of the washers).
2. Everything but the weight and the stretching distance; for example, the setup and other materials used.
3. The more washers added (weight), the more the rubber band stretched.
4. A linear relationship means that as one value increases, so does the other.
5. Answers will vary.

Follow the Bouncing Ball (p. 97)

1. Answers will vary.
2. The bounce chamber was constructed to contain the experiment and to simplify the process of collecting data.
3. Possible variables include an unevenly-shaped ball and a misaligned chamber.
4. A ball compresses when it comes into contact with a surface (the amount it compresses depends on it's velocity), and then expands back to its original shape, reversing the direction of the force imparted upon it by gravity.

The Tallest Tower (p. 99)

Answers will vary.

The Strongest Bridge (p. 102)

Answers will vary.

Oops! The Egg Dropped (p. 104)

Answers will vary.

First-Rate Flying (p. 106)

Answers will vary.